Alluvial Diamond Resource Potential and Production Capacity Assessment of Guinea

By Peter G. Chirico, Katherine C. Malpeli, Mark Van Bockstael, Mamadou Diaby, Kabinet Cissé, Thierno Amadou Diallo, and Mahmoud Sano

Teams of artisanal miners extract diamondiferous gravel from an alluvial flat in Kérouané Prefecture, Guinea. Photo by Mamadou Diaby, CECIDE.

Prepared in cooperation with the Ministère des Mines et de la Géologie of Guinea under the auspices of the U.S. Department of State

Scientific Investigations Report 2012–5256

U.S. Department of the Interior
U.S. Geological Survey

U.S. Department of the Interior
KEN SALAZAR, Secretary

U.S. Geological Survey
Marcia K. McNutt, Director

U.S. Geological Survey, Reston, Virginia: 2012

For more information on the USGS—the Federal source for science about the Earth, its natural and living resources, natural hazards, and the environment, visit http://www.usgs.gov or call 1–888–ASK–USGS.

For an overview of USGS information products, including maps, imagery, and publications, visit http://www.usgs.gov/pubprod

To order this and other USGS information products, visit http://store.usgs.gov

Suggested citation:
Chirico, P.G., Malpeli, K.C., Van Bockstael, Mark, Diaby, Mamadou, Cissé, Kabinet, Diallo, T.A., and Sano, Mahmoud, 2012, Alluvial diamond resource potential and production capacity assessment of Guinea: U.S. Geological Survey Scientific Investigations Report 2012–5256, 49 p. (Available online at http://pubs.usgs.gov/sir/2012/5256/)

Contents

Figures

Tables

Conversion Factors

Multiply	by	To obtain
centimeter (cm)	0.3937	inch (in.)
meter (m)	3.281	foot (ft)
kilometer (km)	0.6214	mile (mi)
square meter (m²)	0.0002471	acre
hectare (ha)	2.471	acre
square kilometer (km²)	247.1	acre
cubic meter (m³)	35.31	cubic foot (ft³)

Temperature in degrees Celsius (°C) can be converted to degrees Fahrenheit (°F):
$$°F = (1.8 \times °C) + 32$$

Abbreviations

AD	Administrative Decision
ALOS	Advanced Land Observing Satellite
AREDOR	Association pour la Recherche de Diamants et d'Or
BEKIMA	Beyla-Kissidougou-Macenta
BEYLA	Société Minière de Beyla
BGR	Bundesanstalt für Geowissenschaften und Rohstoffe
BNE	Bureau National d'Expertise
BRGM	Bureau de Recherches Géologiques et Minières
CECIDE	Centre du Commerce International pour le Développement
CONADOG	Confédération Nationale des Diamantaires et Orpailleurs de Guinée
DEM	digital elevation model
EGED	Entreprise Guinéenne d'Exploitation du Diamant
ETM	Enhanced Thematic Mapper
GIS	geographic information system
GMT	Greenwich Mean Time
GPS	global positioning system
JAXA	Japanese Aerospace Exploration Agency
kt	carat
KPCS	Kimberley Process Certification Scheme
MMEH	Ministère des Mines de l'Energie et de l'Hydraulique
MMG	Ministère des Mines et de la Géologie
PAC	Partnership Africa Canada
PRADD	Property Rights and Artisanal Development Pilot Program
RESTEC	Remote Sensing Technology Center
SNED	Service National d'Exploitation du Diamant
SOGUINEX	Société Guinéenne de Recherches et d'Exploitation Minière
SRRI	Société Real Rock International
USAID	U.S. Agency for International Development
USGS	U.S. Geological Survey
UTM	Universal Transverse Mercator
WGDE	Working Group of Diamond Experts

Alluvial Diamond Resource Potential and Production Capacity Assessment of Guinea

By Peter G. Chirico,[1] Katherine C. Malpeli,[1] Mark Van Bockstael,[2] Mamadou Diaby,[3] Kabinet Cissé,[4] Thierno Amadou Diallo[5], and Mahmoud Sano[6]

Abstract

In May of 2000, a meeting was convened in Kimberley, South Africa, by representatives of the diamond industry and leaders of African governments to develop a certification process intended to assure that export shipments of rough diamonds were free of conflict concerns. Outcomes of the meeting were formally supported later in December of 2000 by the United Nations in a resolution adopted by the General Assembly. By 2002, the Kimberley Process Certification Scheme (KPCS) was ratified and signed by diamond-producing and diamond-importing countries.

The goal of this study was to estimate the alluvial diamond resource endowment and the current production capacity of the alluvial diamond mining sector of Guinea. A modified volume and grade methodology was used to estimate the remaining diamond reserves within Guinea's diamond-iferous regions, while the diamond-production capacity of these zones was estimated by inputting the number of artisanal miners, the number of days artisans work per year, and the average grade of the deposits into a formulaic expression.

Guinea's resource potential was estimated to be approximately 40 million carats, while the production capacity was estimated to lie within a range of 480,000 to 720,000 carats per year. While preliminary results have been produced by integrating historical documents, five fieldwork campaigns, and remote sensing and GIS analysis, significant data gaps remain. The artisanal mining sector is dynamic and is affected by a variety of internal and external factors. Estimates of the number of artisans and deposit variables, such as grade, vary from site to site and from zone to zone. This report has been developed on the basis of the most detailed information available at this time. However, continued fieldwork and evaluation of artisa-nally mined deposits would increase the accuracy of the results.

Introduction

Kimberley Process Certification Scheme

In May of 2000, a meeting was convened in Kimberley, South Africa, by representatives of the diamond industry and leaders of African governments to develop a certification process intended to assure that export shipments of rough diamonds were free of conflict concerns. Outcomes of the meeting were formally supported later in December of 2000 by the United Nations in a resolution adopted by the General Assembly (A/RES/55/56). By 2002, the Kimberley Process Certification Scheme (KPCS) was ratified and signed by diamond-producing and diamond-importing countries.

The KPCS is an international activity whose goal is to prevent the trade of "conflict diamonds" while helping to protect legitimate trade through monitoring the production, exportation, and importation of rough diamonds throughout the world. To accomplish this task, the KPCS requires that every country set up an internal system of controls and checks to prevent conflict diamonds from entering an imported or exported shipment of rough diamonds. Every rough diamond and diamond shipment must be accompanied by a Kimberley Process (KP) certificate and be contained in tamper-proof packaging. The certificate includes an export origin section, an import verification section, and a security slip. The KP also requires that no diamonds are imported from or exported to a nonmember of the KPCS.

---

[1] U.S. Geological Survey.

[2] Chief Officer of International Affairs at the Antwerp World Diamond Centre, Chairman, Technical Committee, World Diamond Council, Chairman of the Working Group of Diamond Experts, the Kimberley Process.

[3] Mining Engineer, Coordinator of the Programme Exploitation Artisanale de Diamant and the Kimberley Process at the Centre du Commerce International pour le Développement (CECIDE).

[4] Executive Director of the Centre du Commerce International pour le Développement (CECIDE).

[5] Geologist, Executive Director General of the Brigade Antifraude des Matières Précieuses.

[6] Mining Engineer, Chef de Section Exploitation Artisanale Diamant, Direction Nationale des Mines, MMG.

Countries that are members of the scheme are required to report the official amount of diamond imports and exports, as well as their value each year to the KP. These data are then made public and provided to other nongovernmental organizations in order to monitor the official statistics reported by all KP members.

Kimberley Process Verification

It is often difficult to obtain independent verification of the diamond-production statistics that are provided by the countries involved in KPCS compliance issues. However, some degree of independent verification can be obtained through an understanding of a country's naturally occurring diamond resources and diamond-production capacity. Studies that integrate these two components can produce a range of estimated values for a country's diamond production, and these estimates can then be compared with the official production statistics released to the KP by the country.

In 2006, the Bureau de Recherches Géologiques et Minières (BRGM) released the first such assessment for the Republic of the Congo. Two methods, integrating measurements of the volume of alluvium based on drainage system models and researching historical data, were used to calculate the alluvial diamond resource within four diamond-bearing zones. A methodology was also implemented for calculating annual production capacity, based on the amount of gravel dug per person per day, gravel grade, the number of active miners, and the number of days miners work per year (Barthélémy and others, 2006). The U.S. Geological Survey (USGS) collaborated with BRGM scientists to produce

subsequent assessments of diamond deposits in Mali and the Central African Republic (Chirico and others, 2010a; 2010b), following the BRGM methodology. The USGS then conducted an assessment of Ghana's diamond deposits, modifying the BRGM methodology by analyzing the deposits at the watershed level and incorporating a geomorphic modeling technique for determining the volume of alluvium (Chirico and others, 2010c).

The goal of this study is to conduct an alluvial diamond resource potential and production capacity assessment for Guinea. This assessment builds on the evolving methodology first introduced by the BRGM and evaluates Guinea's diamond deposits within all diamondiferous subbasins located within five identified diamondiferous regions: Forécariah-Coyah, Kindia-Télimélé, Kissidougou-Macenta, Kérouané, and "Other."

2009 Kimberley Process Administrative Decision on Guinea

During the Seventh Annual Plenary Session of the KPCS, held in Swakopmund, Namibia, in November 2009, an Administrative Decision (AD) pertaining to Guinea was adopted. The Swakopmund AD on Guinea was the result of concerns relating to Guinea's exponential increase in diamond production in the years 2007 and 2008 (table 1). The AD requested that Guinean authorities relaunch a system of internal controls, which include stopping any exports of rough diamonds of suspicious origin. With the passing of the AD, the plenary group agreed that further efforts should be made to assess Guinea's diamond-production capacity (KPCS, 2009).

The Kogbéla artisanal diamond mining site in the Macenta Prefecture, Guinea. Photo by Mamadou Diaby, CECIDE.

Table 1. Annual diamond production in Guinea from the first diamond discovery in 1932 to 2011.

[*, export data; na, not available; —, unknown]

Year	Carats	Year	Carats	Year	Carats
1932	—	1959	657,000*	1986	na
1933	—	1960	1,116,000*	1987	na
1934	0	1961	1,100,000	1988	na
1935	600	1962	na	1989	na
1936	18,897	1963	na	1990	127,000
1937	58,000	1964	na	1991	97,000
1938	61,929	1965	na	1992	95,000
1939	56,316	1966	na	1993	100,000
1940	65,709	1967	na	1994	100,000
1941	57,736	1968	na	1995	452,019
1942	49,866	1969	na	1996	364,013
1943	36,193	1970	na	1997	379,639
1944	69,726	1971	na	1998	355,011
1945	72,802	1972	na	1999	357,386
1946	51,830	1973	na	2000	327,000
1947	62,310	1974	na	2001	364,000
1948	89,490	1975	na	2002	491,000
1949	na	1976	na	2003	666,000
1950	na	1977	na	2004	673,893
1951	na	1978	na	2005	548,522
1952	na	1979	na	2006	473,862
1953	na	1980	na	2007	1,018,723
1954	144,490	1981	na	2008	3,098,490
1955	210,640	1982	na	2009	696,731
1956	256,816	1983	na	2010	374,096
1957	92,194	1984	na	2011	303,785
1958	118,500	1985	na		

Data sources: 1934–1948, 1954–1958 (Bardet, 1974); 1959–1961 (Moyar and Buxtant, 1963); 1990–1994 (Izon, 1994); 1995–1999 (Swiecki, 2008); 2002–2003 (Bermúdez-Lugo, 2004); 2004–2011 (KP Statistics).

Historical Setting

The History of Diamond Mining in Guinea

Diamonds were discovered in Guinea in 1932, in alluvial deposits of the Upper Makona Valley of the Macenta Prefecture. Commercial diamond mining was initiated in 1935 near Baradou, in Kissidougou, by the Société Guinéenne de Recherches et d'Exploitation Minière (SOGUINEX) (Rombouts, 1987a). By 1939, there were active mines near Baradou, Fenaria, Faradou, and Banankoro (PRADD, 2008). In 1952, the first kimberlitic body was discovered by H. Haggard, and additional pipes and dikes were discovered by Russian geologists from 1963 to 1967 (Rombouts, 1987a). In 1953, the Société Minière de Beyla (BEYLA) began mining at Bounoudou, and from 1956 to 1960 Beyla-Kissidougou-Macenta (BEKIMA) actively mined the region. Guinea gained independence from France in 1958. In 1961, state-owned Entreprise Guinéenne d'Exploitation du Diamant (EGED) took over mining operations from BEKIMA and BEYLA (Greenhalgh, 1985). The company eventually shut down in 1973. Figure 1 summarizes events in the history of industrial and alluvial diamond mining in Guinea.

It is probable that artisanal diamond mining began in conjunction with commercial mining activities. Beginning in 1956, an agreement was made between SOGUINEX and the Guinean government, in which SOGUINEX ceded a portion of their territory to be used for artisanal mining. The agreement was meant to reduce the high number of illicit miners crossing the border from Sierra Leone, during a crackdown on artisanal mining in that country (Greenhalgh, 1985). However, the agreement failed after several years. By 1959, there were an estimated 41,000 artisanal diggers mining in Guinea, approximately half of whom were registered and many of whom had come from Sierra Leone. As a result, diamond exports increased rapidly from 118,500 carats (kt) in 1959 to over 1 million kt by 1961. That same year, the government nationalized all foreign company assets, including $1.5 million worth of diamonds stored in the SOGUINEX vault in Guinea (Janse, 1996). The combination of the nationalization of assets and the influx of displaced miners from Sierra Leone contributed to the high diamond exports from 1958 through 1962 (table 1). In 1981, the Association pour la Recherche de Diamants et d'Or (AREDOR), a partnership between the Guinean government and the Australian company Bridge Oil, was developed. AREDOR began industrial alluvial mining outside of Kérouané along the Baoulé River near Gbenko in 1984. Initial exploration indicated very high value deposits, but subsequent mining showed erratic distribution and fluctuations in stone size, and by 1994, AREDOR had closed (Morgan and others, 1992). However, that same year, Guinea was ranked number 10 in the world in the production of diamonds by volume and 9 in production by value, with the average value per carat at $300 (Janse, 1996). Between 1980 and 1984, artisanal mining was authorized once again and

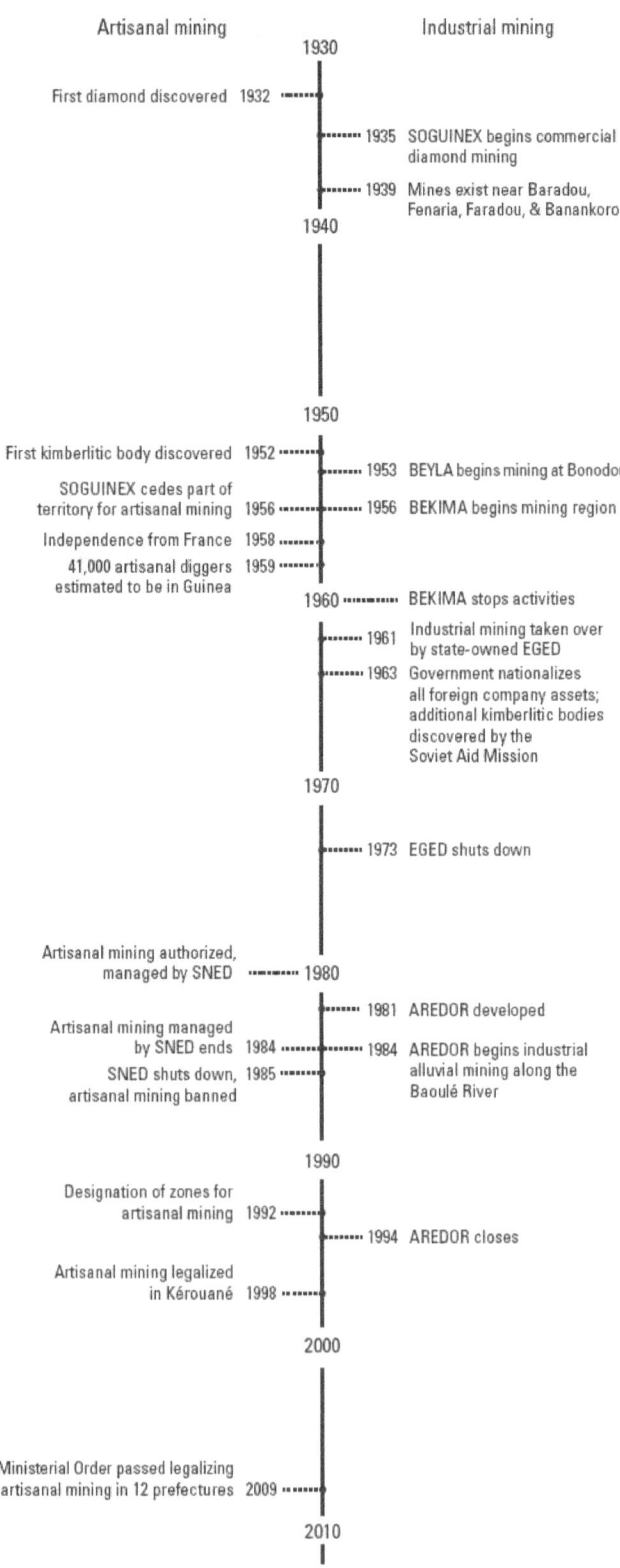

gu e 1. Events in the history of artisanal and industrial mining in Guinea.

this time was managed by a government agency known as the Service National d'Exploitation du Diamant (SNED). Led by a change in government mining policies, SNED shut down in 1985, and artisanal diamond mining was banned. Levels of illegal artisanal mining activities subsequently increased. A third attempt to legalize artisanal diamond mining was made in 1992, with the designation of zones in which artisanal diamond mining was authorized. The main zone was within the AREDOR concession. In 1998, artisanal mining was legalized in the Kérouané Prefecture, on the former AREDOR concession property. In 2009, the Guinean government again sought to legalize artisanal diamond production by passing the 2009 Ministerial Order.

Guinea's Mining Code and 2009 Legislation

The Ministère des Mines et de la Géologie (MMG) is the main governmental organization charged with regulating Guinea's mining industry. In 2009, a Ministerial Order was passed which legalized artisanal diamond mining in particular zones within 11 prefectures. These prefectures include Kindia, Télimélé, Forécariah, Coyah, Kérouané, Kissidougou, Guéckédou, Macenta, N'Zérékoré, Beyla, and Yomou (fig. 2). A twelfth prefecture, Boffa, was also authorized for artisanal mining, but this zone is legalized for the mining of rubies and other precious stones (Ministère des Mines de l'Energie et de l'Hydraulique, MMEH, 2009). This Ministerial Order is the largest expansion of legal artisanal mining in Guinea's history.

Guinea's 1995 Mining Code provides the current legislative framework for mining activities. Within this code, artisanal operations are defined as activities that consist of small-scale operations in which nonmechanized methods are employed.

Artisanal parcels must be no larger than 1–2 hectares in size and are limited to 50 employees per site. Guinean nationals are awarded an Artisanal Operation License by the MMG. This license permits mining within a defined perimeter located in a designated artisanal mining zone. It is valid for a period of 1 year but can be renewed (PRADD, 2008).

The legal flow of diamonds involves only certified and registered miners, collectors, and stakeholders (fig. 3). The artisanal operation title holder sells the stones mined at his site to either *agents collecteurs* or to *comptoirs d'achat*. The *agents collecteurs* are authorized by the MMG to buy and sell diamonds from artisanal operations and can only sell stones to the *comptoirs d'achat*. The *comptoirs d'achat* are authorized by the MMG to buy, import, and export diamonds mined in artisanal operations. The *comptoirs d'achat* then sell to the *Bureau National d'Expertise* (BNE) (Republic of Guinea, 1995). The BNE was established in 1993 and is responsible for the marketing of artisanal diamonds. It is also Guinea's precious metal and gem evaluation agency and verifies the origin of diamonds and determines the export value and tax obligation of diamond exports (PRADD, 2008).

Previous Estimate of Diamond Resources

A thorough review of the mineral economy of Guinea was conducted by Morgan and others (1992) and provides information on deposits and estimated reserves of many of the key minerals found in Guinea. Total diamond resources in Guinea were estimated to be 25–30 million carats (Morgan and others, 1992). These estimates focused mainly on the known deposits within the Kérouané, Beyla, Macenta, Guéckédou, and Kissidougou Prefectures.

The Wandadou artisanal diamond mining site in the Kérouané Prefecture, Guinea. Photo by Mamadou Diaby, CECIDE.

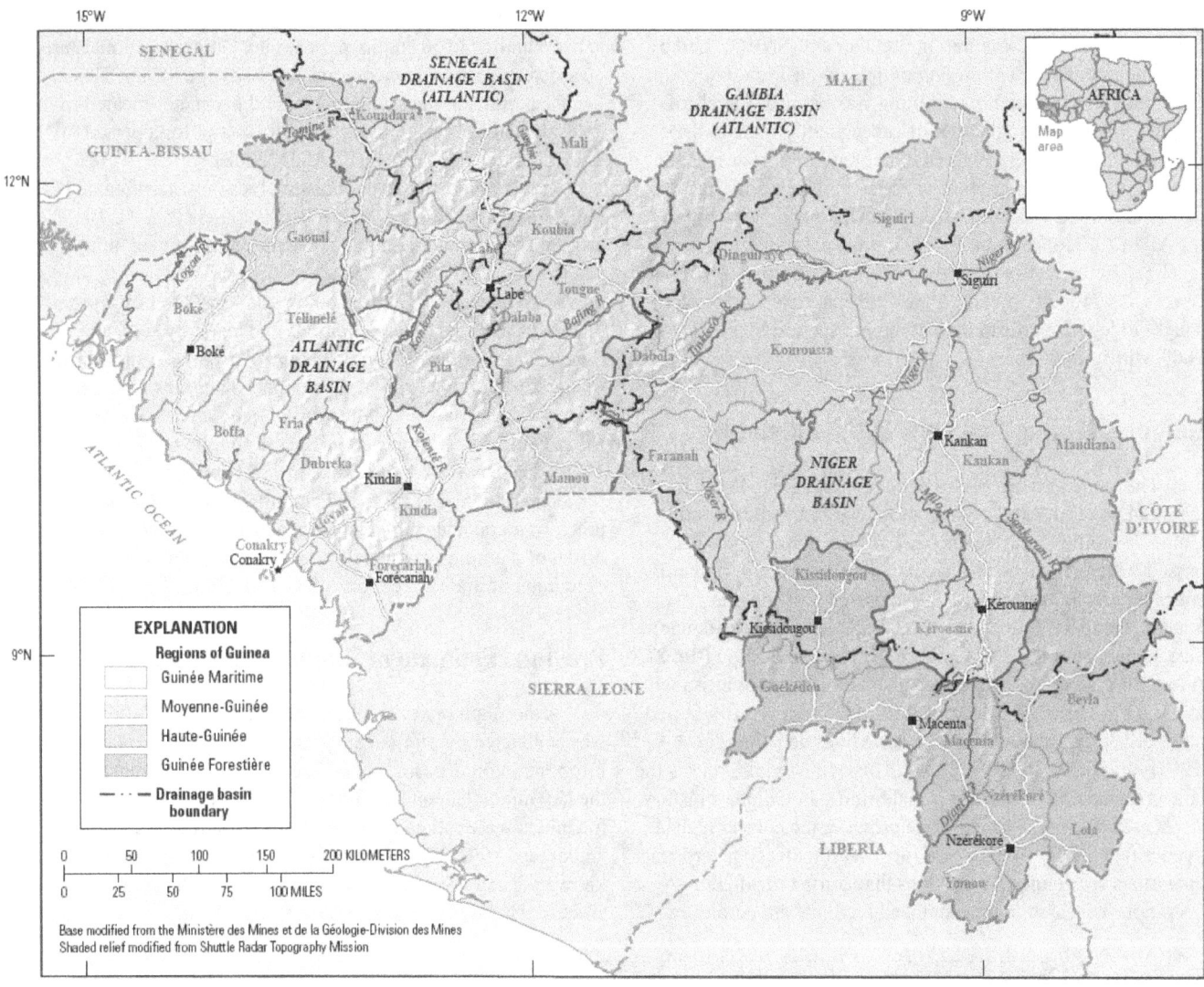

Figure 2.   The general geography of Guinea.

Geologic Setting

Geographic Setting

Guinea is on the Atlantic coast of West Africa and has an area of 245,857 square kilometers (km²). The country's four geographic regions are a narrow coastal belt running north to south (Guinée Maritime), a pastoral highlands (Moyenne-Guinée), a northeastern savannah (Haute-Guinée), and the southeastern rainforest (Guinée Forestière). The coastal plain of the west is mostly flat. East of the coastal area are hills and mountains of the interior (fig. 2). Positioned between latitudes 7° and 13° N., Guinea's climate is tropical, with a monsoonal rainy season from April to November. Across the country, average annual rainfall amounts range from 1,500 millimeters (mm) to 4,300 mm. The western part of the country receives

more rain than the eastern part. The temperature range is 22–25 degrees Celsius (°C) in the wet season and 25–27 °C in the dry season (Bering and others, 1998).

Guinea is a mineral-rich nation. It has the world's largest bauxite reserves, as well as substantial reserves of iron ore, gold, uranium, and diamonds. Diamond deposits and kimberlites are located mainly in Guinée Forestière and the southern half of Haute-Guinée, with other occurrences located in the west near Conakry. Haute-Guinée is divided between the Niger River drainage basin, which has its source in Guinea's highlands, and the Atlantic drainage basin (fig. 2). Regional-scale tectonics have caused a south-southeast tilting resulting in the capturing of the headwaters of the Niger basin rivers by Atlantic drainage basin rivers. This has heavily influenced the occurrence of diamond deposits in the region (Rombouts, 1987a).

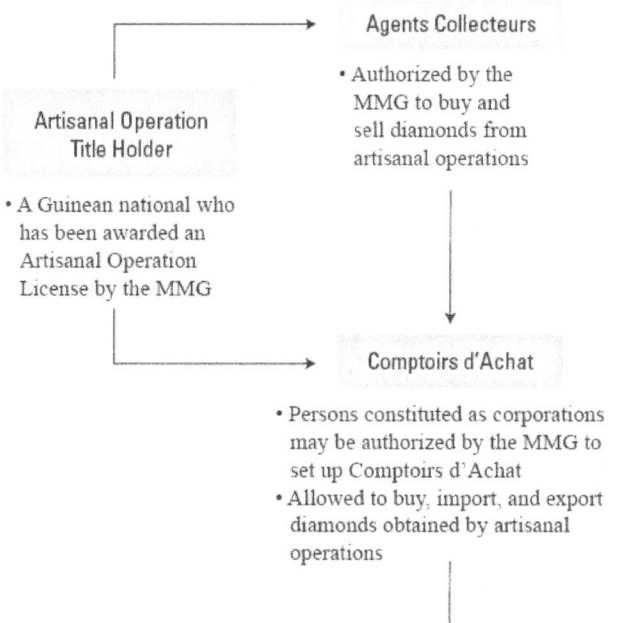

Artisanal Operation Title Holder

• A Guinean national who has been awarded an Artisanal Operation License by the MMG

Agents Collecteurs

• Authorized by the MMG to buy and sell diamonds from artisanal operations

Comptoirs d'Achat

• Persons constituted as corporations may be authorized by the MMG to set up Comptoirs d'Achat
• Allowed to buy, import, and export diamonds obtained by artisanal operations

Bureau National d'Expertise (BNE)

• Manages the marketing of artisanal diamond mining
• Determines the export value and tax obligation
• Verifies the origin of the diamonds

Figure 3. Diagram showing the flow of artisanally mined diamonds in Guinea. (MMG, Ministère des Mines et de la Géologie)

The Geology and Diamond Deposits of Guinea

Northern Guinea is made up of Neoproterozoic and Paleozoic sediments composed of a basal tillite and sandstones, marls, and quartzites. Mesozoic dolerites and gabbroic dunites and Paleoproterozoic micha schists are also present (fig. 4). The eastern two-thirds of the country is dominated by the gneiss, migmatites, granitoids, and schists of the Kenema-Man domain and the Guinea Rise, except for a portion in the northeast, which is composed of Paleoproterozoic mica schists of the Birimian System. Along the coast, there is a thin strip of Cenozoic marine and fluvial sediments (Schlüter and Trauth, 2008).

Guinea's diamond deposits can be subdivided into primary diamond occurrences and alluvial deposits. The primary diamond deposits occur in kimberlitic pipes, dikes, and fractures. Nineteen pipes and numerous dikes and fractures have been identified in southeastern Guinea. The majority were discovered by the Soviets between 1963 and 1967. All of the known kimberlites appear to be diamondiferous, with the dikes generally having higher grades than the pipes (Rombouts, 1987a). Guinea's known primary diamond occurrences are located in the Kenema-Man domain of the West African Craton, which is underlain by granitoid rocks of Archean and Lower Proterozoic age. The basement rocks are chiefly granitic; however, Archean gneiss and quartz and Precambrian doleritic dikes and sills also occur (fig. 5).

Alluvial Diamond Deposits

Regional tectonic uplift as well as climatic events over the past 40,000 years have shaped the fluvial geomorphology of the region and the location of alluvial diamond deposits (Rombouts, 1987a; Teeuw and others, 1991; Sutherland, 1993). Two types of alluvial diamond deposits occur in Guinea, based primarily on the distance from their source rocks (Bardet, 1974). The first type consists of eluvial-alluvial deposits, which are found near the source rocks. The second type is the alluvial deposit, which has been transported far from its source. In the first case, rich to very rich diamond deposits are found in the alluvial flats of small tributaries. The second type consists of reworked gravels, with the richest deposits located in the alluvial flats and low terraces (Bardet, 1974).

The dispersal of diamond deposits and the size and quality of the diamonds in Guinea have been influenced by the fluvial geomorphology of the Atlantic and Niger drainage basins (fig. 6). The young relief and incised channels of the Atlantic drainage have caused alluvial sediments to be deposited on older terraces, with reworking of the gravels. As a result, the alluvial flats within this basin tend to contain higher grade diamonds than those found in the terraces. In contrast, the Niger River drainage relief is mature, with broad valleys and well-developed terraces and flats. The more recent alluvial materials do not regularly superimpose themselves on older terraces, and therefore the alluvial flat deposits are not necessarily richer than the terraces (Rombouts, 1987a).

Previous Studies Examining Guinea's Alluvial Diamond Deposits

Several studies have been conducted on the morphological evolution of Guinea's diamond deposits, mainly within the rich southeastern diamond zone. While many of these studies are conducted at the subwatershed scale, they document the geomorphology, location, and richness of the evaluated deposits, providing insight related to the characteristics and depositional histories of Guinean diamonds and assisting with the modeling of other alluvial deposits. Information regarding the location and size of the alluvial flats and terraces and the concentration of diamonds throughout these deposits, over-burden and gravel thickness, and grade data are particularly useful for modeling purposes. Table 2 contains all available soil profile data summarized in the previous studies of Guinea's diamond deposits.

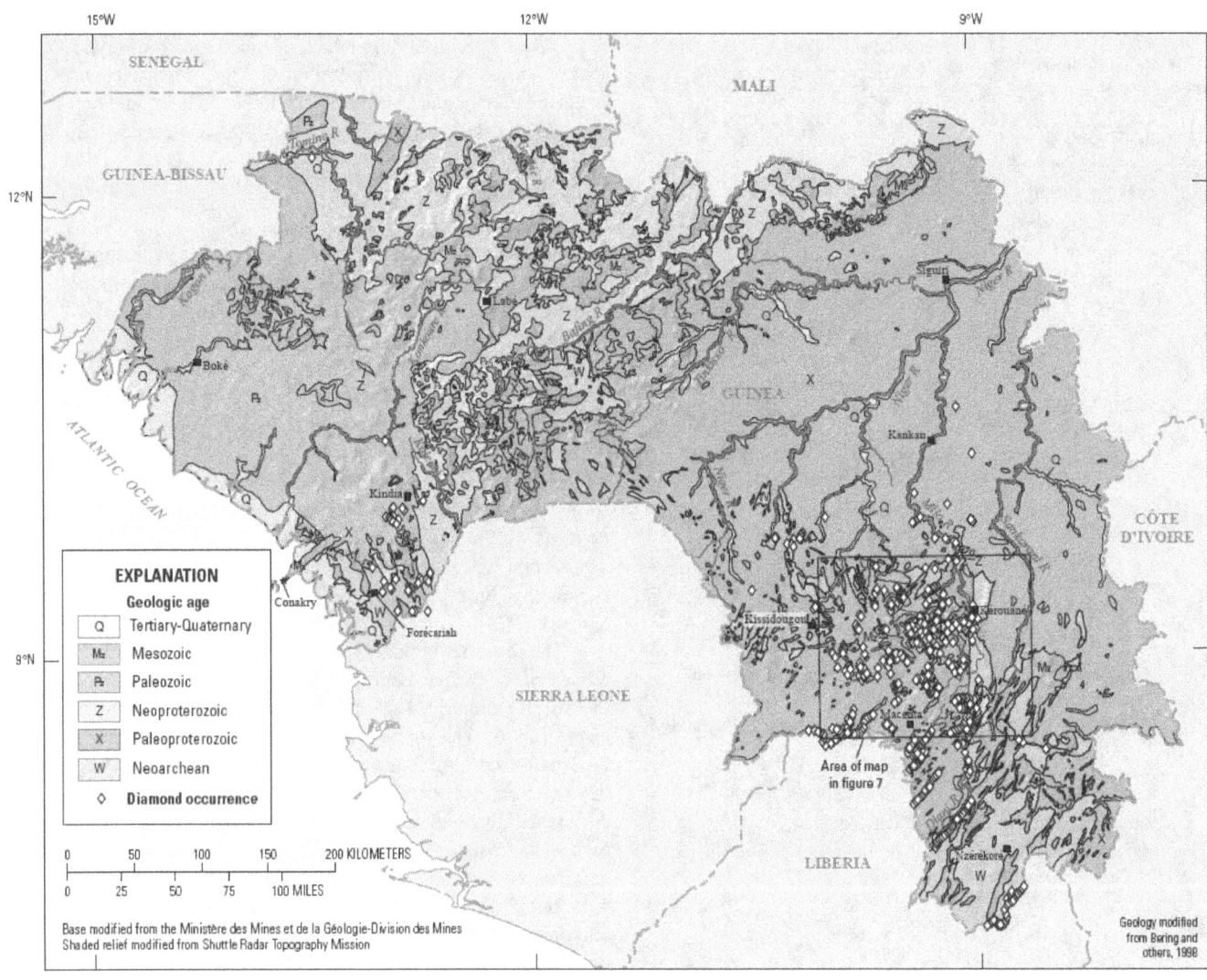

Figure 4.   The general geology of Guinea and diamond deposit occurrences.

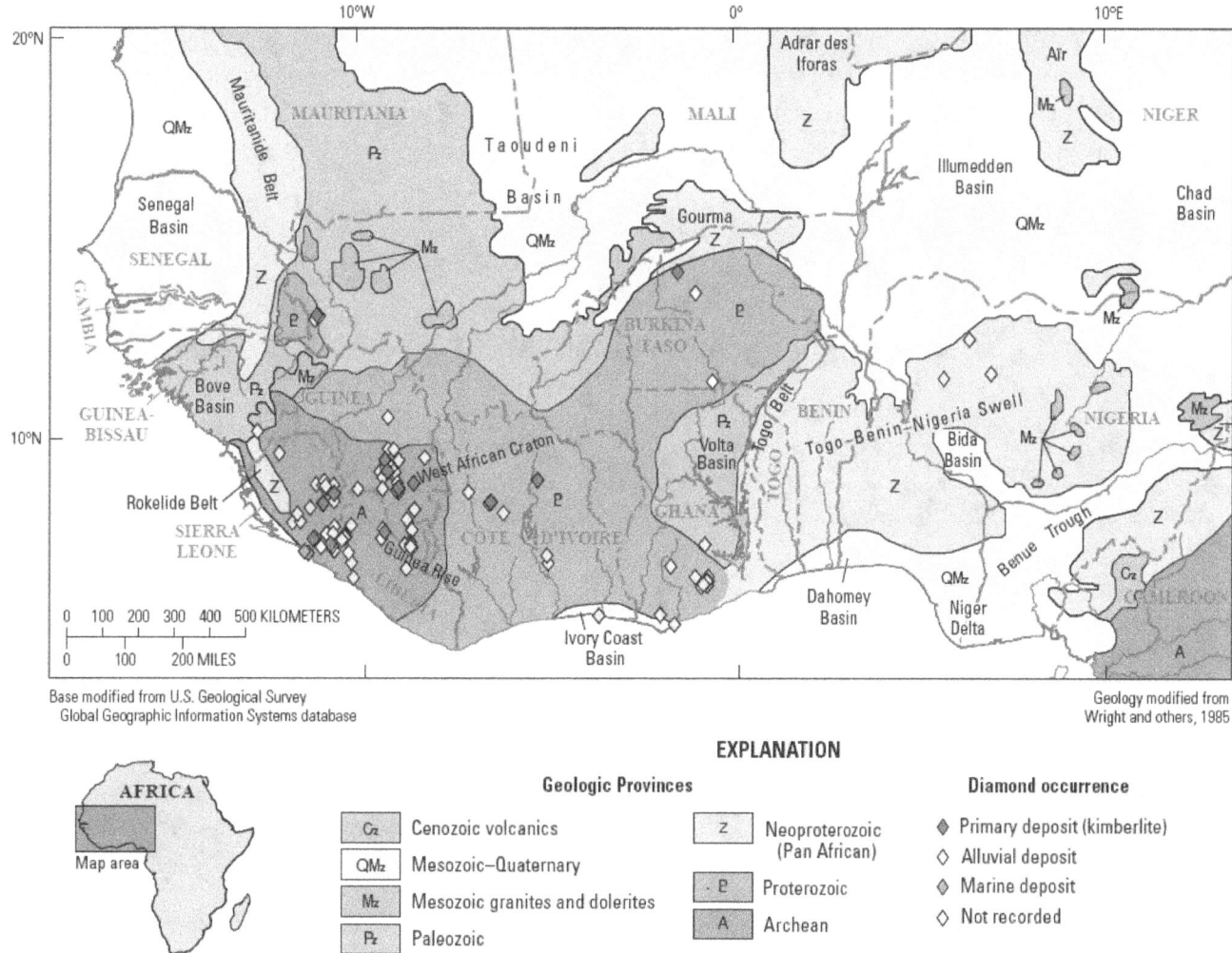

Base modified from U.S. Geological Survey
  Global Geographic Information Systems database

Geology modified from
Wright and others, 1985

EXPLANATION

AFRICA

Map area

Geologic Provinces

Cz	Cenozoic volcanics	z	Neoproterozoic (Pan African)
QMz	Mesozoic–Quaternary	P	Proterozoic
Mz	Mesozoic granites and dolerites	A	Archean
Pz	Paleozoic		

Diamond occurrence

◈ Primary deposit (kimberlite)
◇ Alluvial deposit
◈ Marine deposit
◇ Not recorded

Figure 5.   The geologic provinces and diamond locations of West Africa.

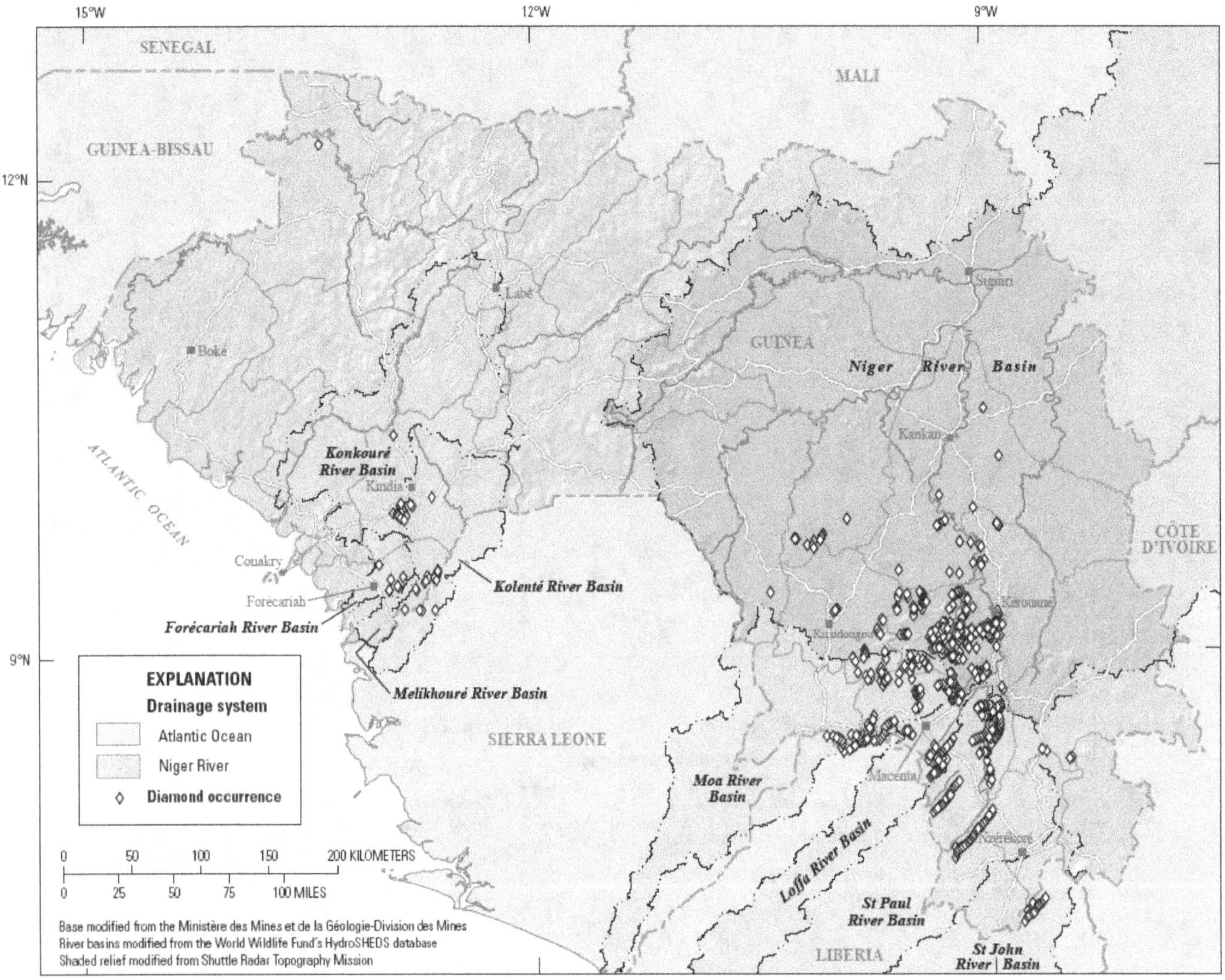

Figure 6.   The Niger River and Atlantic drainage basin watersheds in Guinea.

Table 2.   Soil profile data obtained from previous studies in Guinea.

[m, meter; —, unknown]

Author	Location	Drainage basin	Geomorphology	Gravel thickness (m)	Overburden thickness (m)	Overburden description
Rombouts (1987a)	Baoulé River	Niger	Alluvial flat	0.3–0.45	3–8	1–3 m grit and sand; 2–5 m silt and clay
Rombouts (1987a)	Baoulé River	Niger	Low terrace	—	0.2–10	—
Sutherland and Robinson (1996)	Sarabaya River	Niger	Alluvial flat and low terrace	0.34	3.8	Sand and clay, capped with brown clay
Sutherland and Robinson (1996)	Sarabaya River	Niger	High terrace	0.3	—	—
Stellar (2010)	Mandala River	Atlantic	Alluvial flat	0.7	4.5	Layers of sand and clay
Bering and others (1998)	Diana River near Bounoudou	Atlantic	Alluvial flat	0.3	5–7	Whitish sand and red gravelly sand
Bering and others (1998)	Baoulé River near Gbenko	Niger	Alluvial flat	0.2–0.3	6	Silt and sand

The AREDOR concession in southeastern Guinea remains the most explored and best documented alluvial deposit in the country. The alluvial flats of this deposit range from 250 meters (m) to 1 kilometer (km) wide, with the thickness of alluvial overburden ranging from 4 to 6 m and gravel thickness ranging from 0.30 m to 0.45 m (Rombouts, 1987a). Sutherland (1993) conducted a study of three drainage basins upstream of the AREDOR concession: the Mandala, Bouloumba, and Baoulé. The upper Bouloumba and the upper Baoulé are within the Niger River drainage basin, and the Mandala flows into the Atlantic via the Makona River (fig. 7). The large streams within each basin consist of flood plains and low terraces, with the low terrace deposits lying 1–2 m above the flood plain. Combined, these two features vary in width from 10 to 50 m along tributaries and reach up to 500 m in width along the trunk streams.

Sutherland (1993) noted that in the Mandala basin, the highest diamond concentration occurs along trunk stream valleys, while lower concentrations occur in tributary streams. This is due to the headwater extension of the Atlantic drainage system, which resulted in capturing and downcutting the Mandala River's main channel. Subsequent rejuvenation of the basin flushed tributaries into the trunk stream, increasing the trunk's sediment volumes. In contrast, the Bouloumba and Baoulé basins have experienced reduced discharge and are therefore less likely to transport heavy minerals from their sources. In both basins, the tributary streams contain the highest concentrations of diamonds, while the trunk streams contain low to moderate concentrations (Sutherland, 1993).

The Sarabaya River, located within the upper Bouloumba drainage basin, contains flood plain and low terrace and high terrace deposits (Sutherland and Robinson, 1996).

The concentration of diamond deposits in the Sarabaya is controlled by the evolution of the Bouloumba drainage basin. The capturing of the southern portion of the Bouloumba basin by the Milo basin may have resulted in a greater retention of sediments in tributaries of the Bouloumba, such as the Sarabaya. The highest concentrations of diamonds within the river are found in the valley bottom. Diamonds were also noted to have occurred in the gravels in the upper terrace of the northern valley side (Sutherland and Robinson, 1996).

In 2004, Sutherland conducted resource estimates for the Mandala and Ouria permits owned by Stellar Diamonds Limited (2010), in the Kérouané Prefecture. He concluded that 547,000 kt remain in the Mandala and N'Kéléyani valley gravels, with the average gravel grades of zones within this area ranging from 0.2 to 0.63 kt/m². Furthermore, Sutherland (1993) estimated 144,000 kt remaining in the Ouria concession, with an average gravel grade of 0.80 kt/m² (Walker and Sobie, 2008). It was also noted that Mandala's alluvial diamond deposits have a relatively high grade. Alluvial flats and low terraces in the Mandala watershed are often diamondiferous, while the high terraces do not contain deposits of economic interest. The Ouria diamonds are smaller than those recovered in Mandala, but they are of similar quality (Walker and Sobie, 2008).

Bering and others (1998) characterize the alluvium of two regions in southeastern Guinea, one flowing into the Atlantic drainage basin and the other into the Niger drainage basin. Along the Diani River, near Bounoudou, the overburden is noted to be 5 to 7 m thick and consists of a layer of white sand overlying 1 to 2 m of red gravelly sand. The diamondiferous gravel layer is 0.3 m thick and lies atop Archean gneiss. This sequence is a typical profile for alluvial deposits within the

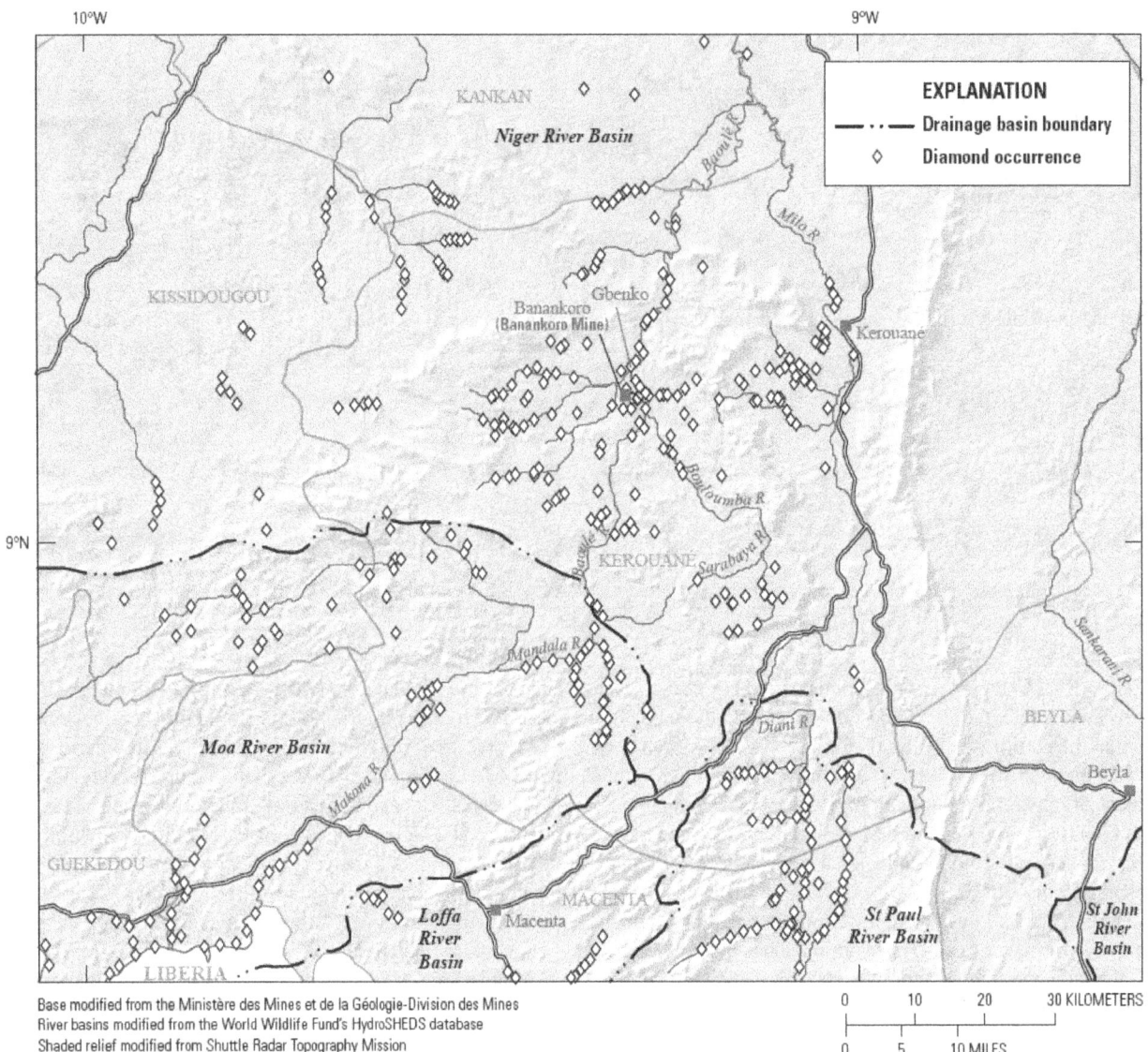

Base modified from the Ministère des Mines et de la Géologie-Division des Mines
River basins modified from the World Wildlife Fund's HydroSHEDS database
Shaded relief modified from Shuttle Radar Topography Mission

Figure 7.   The locations of diamondiferous rivers within the Mandala, Bouloumba, and Baoulé drainage basins examined by Sutherland (1993) and Rombouts (1987a).

Atlantic drainage basin, where the fluvial systems' high energy causes deep erosion. The soil profile for the second area, near Gbenko, which drains north to the Baoulé River and is part of the Niger system, consists of 6 m of silt and sand on top of a diamondiferous gravel layer that is 0.2 to 0.3 m thick (fig. 7; table 2).

Almost all of the diamond-carrying rivers of the Niger drainage system are found in the Gbenko area in Guinea's Kérouané Prefecture. Diamonds are transported first through the Baoulé and Milo Rivers and then enter the Niger River. The Baoulé River's alluvial flats near Gbenko are between 250 and 1,000 m wide, and the combined width of the flat and terrace deposits can reach 2.5 km. The alluvial flat sediments are generally 4 to 6 m thick, and terrace deposits range from several centimeters to more than 10 m thick. The Gbenko and Banankoro areas appear to have the largest and best crystallized diamonds. Almost all tributaries of the Baoulé River that flow into the Gbenko and Banankoro area have diamonds. Here, the average size of stones greater than 2 mm wide is 0.8 kt. The size varies from 0.1 to 0.5 kt in other Guinean deposits. The Bimboko tributary has the highest concentration of large stones in the Gbenko area.

Study Areas

Guinea's alluvial diamond deposits are largely located downstream from the kimberlitic bodies of the southeast. However, a number of secondary diamond deposits do not appear to be correlated with known kimberlitic source rocks. Such deposits are found in the Kindia, Forécariah, and Télimélé Prefectures in Guinée-Maritime, as well as the Fouta-Djallon region of western Moyenne-Guinée. The geology of these areas consists of Paleozoic sandstone and conglomerate, Mesozoic dolerite, and Archean-Proterozoic mafic gneiss. The alluvial diamond deposits in the Kindia-Télimélé region are formed as paleoplacers in deep bedrock fractures (Bering and others, 1998). Diamond deposits are also found in northeastern Guinea and in the western Kindia region. The origin and age of the former are unclear, though there is speculation that they were transported from the Banankoro region within the Niger hydrologic system, flowing to the northeast (Bering and others, 1998).

The majority of previous studies on the diamond deposits of Guinea are focused on the Kissidougou-Kérouané-Macenta triangle. While Rombouts (1987a) recognized diamond deposits in the Forécariah-Kindia and Fouta-Djalon region, he concluded that no known workable diamond deposits exist outside of southeastern Guinea (Rombouts, 1987a). Four western prefectures, Forécariah, Coyah, Kindia, and Télimélé, were visited over the course of the first three field missions. The more intensively mined southeastern prefectures of Kissidougou, Macenta, and Kérouané were visited during the initial field mission in 2010 and again during the final mission in May 2012 (fig. 8; table 3). The alluvial diamond resource assessment of this study summarizes existing information on known Guinean diamond deposits. These data are then compared with results of recent fieldwork conducted in the lesser studied alluvial deposits of western Guinea.

Fieldwork Missions in Guinea

Several separate field campaigns have been carried out in support of the AD request. To conduct the first of these missions, the USGS partnered with the KP Working Group of Diamond Experts (WGDE) and Guinea's Ministère des Mines et de la Géologie (MMG) to conduct a preliminary field campaign from April 24 to May 2, 2010. The field team was made up of Mark Van Bockstael of the WGDE, Peter Chirico of the USGS, Mamadou Soumanou of the MMG, and a team of mining engineers and geologists from the MMG. Diamond mining sites were visited in western Guinea's Kindia, Forécariah, Coyah, and Télimélé Prefectures, where the Guinean government identified newly discovered deposits mined by artisans. Several mining sites in southeastern Guinea's Kissidougou Prefecture were also visited as part of the study.

As a followup, the USGS conducted a series of technical assistance workshops and trainings in Conakry, Guinea, from May 31 to June 4, 2011. The purpose of the followup work was threefold. First, the results of the preliminary report drafted after the 2010 USGS/WGDE visit were presented to members of Guinea's MMG and representatives of several civil society organizations including the Centre du Commerce International pour le Développement (CECIDE) and the Confédération Nationale des Diamantaires et Orpailleurs de Guinée (CONADOG), with support from Partnership Africa Canada (PAC), the U.S. Embassy (Conakry), and the U.S. Agency for International Development (USAID). Second, the methodology used by the USGS to conduct the diamond resource potential and production capacity assessment of Guinea was presented in detail to the MMG and civil society representatives. This presentation included training on using handheld global positioning system (GPS) units in the field. Third, the USGS demonstrated to members of the MMG and civil society organizations methods for conducting fieldwork and data gathering techniques during a trip to diamond mining sites in the Forécariah Prefecture. During this trip, preliminary plans were made for members of CECIDE and the MMG to carry out joint fieldwork missions in Guinea, following the methodology presented by USGS, to enhance the USGS's database on artisanal diamond mining activities in Guinea.

In November of 2011, CECIDE met with the MMG, presented a plan to move forward, and requested formal cooperation from the Guinean government. Additional discussions occurred in January 2012 to clarify points from the November meeting. The details of the agreement were finalized, and in March of 2012, fieldwork was conducted by a team composed of Mamadou Diaby, Coordinator of the Programme Exploitation Artisanale de Diamant and the Kimberley Process at CECIDE, Thierno Amadou Diallo, Executive Director General of the Brigade Antifraude des Matières Précieuses, and Mahmoud Sano, Chef de Section Exploitation Artisanale Diamant at the MMG. The team was supported locally by guides and representatives of local mining cooperatives. The team visited five artisanal diamond mining sites in the Kindia and Télimélé Prefectures. Later that month, the USGS returned and conducted a joint fieldwork mission with geologists from CECIDE and the MMG and visited seven artisanal diamond mining sites in the Forécariah and Coyah Prefectures.

The final fieldwork mission were conducted in May 2012 by the team. Seventeen artisanal diamond mining sites in the intensively mined Macenta and Kérouané Prefectures of southeastern Guinea were visited during this mission (USAID, 2012). The data collected by the Guinean partners was transferred by CECIDE directly to USGS for analysis. In total, 40 field sites were mapped over the course of the 2010–2012 field missions.

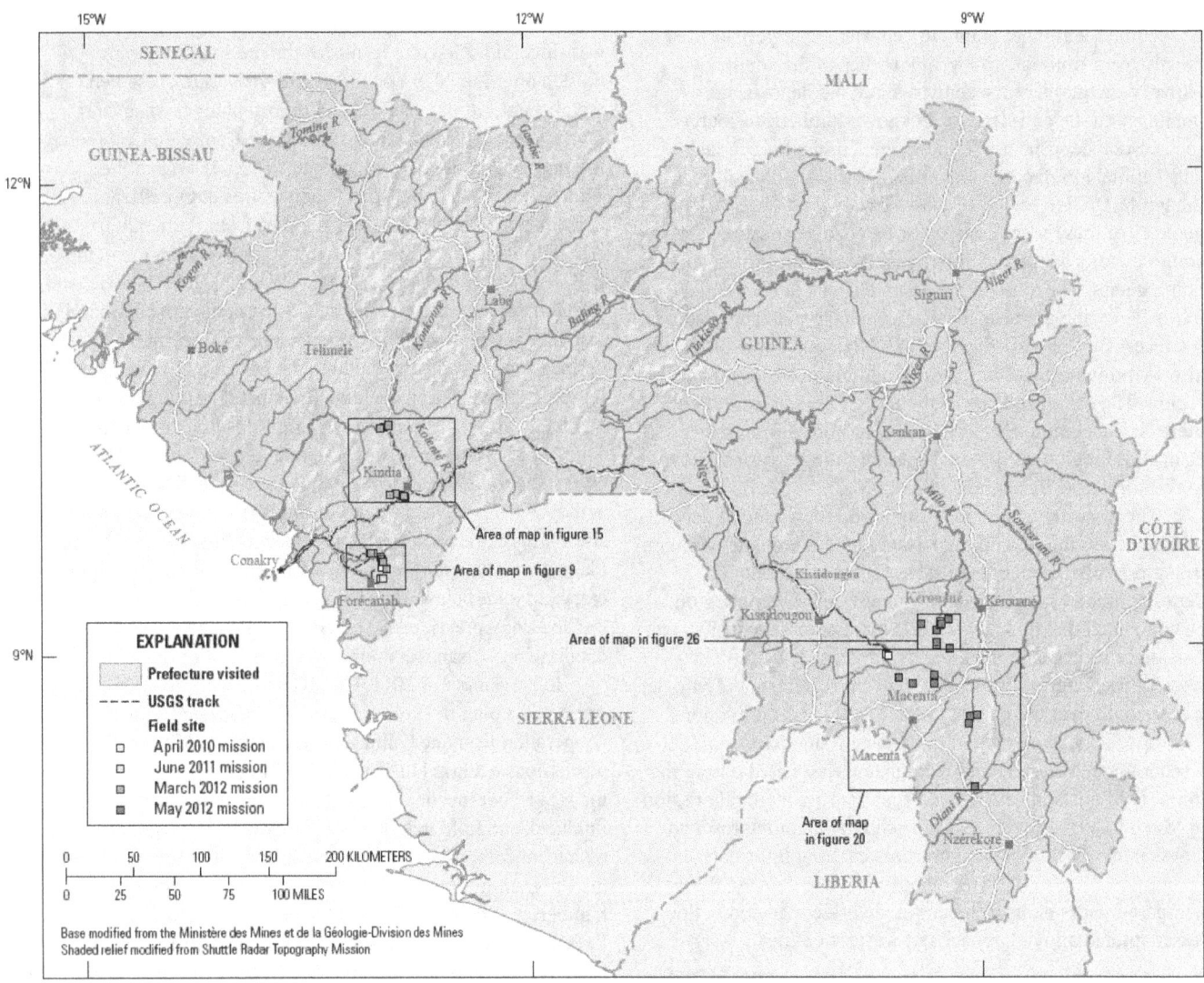

Figure 8.   The field sites visited during fieldwork conducted in Guinea in April 2010, June 2011, and March and May 2012.

Table 3.   General information about each of the sites visited in Guinea from 2010 to 2012.

[m, meter; —, unknown]

Site	Prefecture	Date of fieldwork	Situation during visit	Number of miners	Geomorphic zone	Gravel thickness (m)	Overburden thickness (m)
Forécariah-Coyah Region							
Kourouya 1	Forécariah	4/27/2010	Temporarily inactive	100	Low terrace	0.4	1
Kourouya 2	Forécariah	4/27/2010	Temporarily inactive	—	Alluvial flat	0.2	1.7
Kenenday	Coyah/Kindia	4/27/2010	Active	—	Alluvial flat	—	—
Baschia	Forécariah	4/27/2010	Active	30	Alluvial flat	0.25	1.5
Heremakono	Forécariah	6/3/2011	Active	21	Alluvial flat	0.5	4
Bouramaya	Kindia	3/6/2012	Active	220	Active channel, low terrace	0.5	2.5–9.5
Kenenday North	Coyah/Kindia	3/7/2012	Active	3	Active channel, low terrace	0.2–0.5	4
Sansangui 1	Forécariah	3/7/2012	Active	300–400	Low terrace, ancient terrace	0.2–2	1.5
Sansangui 2	Forécariah	3/7/2012	Active	1,000	Active channel	—	0
Gbomilo	Forécariah	3/7/2012	Active	9	Active channel	0.3–0.5	1.5–4
Banyama	Forécariah	3/8/2012	Active	300–600	Alluvial flat	0.5	1.5–2
Safoulé	Forécariah	3/8/2012	Active	1,200	Active channel, low terrace	0.5–1	1–2
Kindia-Télimélé Region							
Férékouré	Kindia	4/28/2010	Active	—	Paleochannel	4	8
Foulaya	Kindia	4/30/2010	Inactive	—	Paleochannel	0.5	6
Samoriya 1	Kindia	4/30/2010	Inactive	—	Paleochannel	0.5	5
Kafour	Télimélé	3/3/2012	Active	30	Paleochannel, active channel	—	—
Témé	Kindia	3/3/2012	Active	3	Active channel	1	3
Angola	Kindia	3/4/2012	Active	60	Active channel	0.8–1	5–15
Menyima	Kindia	3/4/2012	Active	14	Paleochannel, active channel, low terrace	1	12
Samoriya 2	Kindia	3/4/2012	Active	19	Active channel, low terrace	1	11
Momo Banfouruyu	Kindia	3/4/2012	Active	9	Paleochannel, ancient terrace	1–1.5	4.5–6
Kissidougou-Macenta Region							
Fondiya	Kissidougou	4/29/2010	Active	—	Low terrace	0.5	7.6
Dakoudou	Kissidougou	4/29/2010	Active	—	Alluvial flat	0.4–0.5	3
Gbanwani	Macenta	5/19/2012	Active	800	Low terrace	0.5	5.5–6.5
Societe RRI	Macenta	5/19/2012	Active	—	—	—	—
Kambaya	Macenta	5/20/2012	Active	600	Active channel, low terrace, ancient terrace	0.3	4.7
Kodjan	Macenta	5/20/2012	Active	50	Active channel, low terrace, ancient terrace	0.5–1	1.5
Kolokolo	Macenta	5/20/2012	Active	3	—	—	—
Bayama	Macenta	5/21/2012	Active	30	Active channel, ancient terrace	1	0
Kogbéla	Macenta	5/22/2012	Active	220	Active channel, low terrace, ancient terrace	0.25–0.5	3.75–4.5
Wadagbolofé	Macenta	5/22/2012	Active	200	Low terrace, ancient terrace	0.3–0.5	6.2–6.5
Kérouané Region							
Loia	Kérouané	5/22/2012	Active	20	Alluvial flat	0.3–0.5	5.7–6.5
Finariah	Kérouané	5/22/2012	Active	500	Alluvial flat	0.25–0.5	4.75–5.5
Somassania	Kérouané	5/23/2012	Active	500	Low terrace, ancient terrace	0.5–1	6.5–7
Faraba	Kérouané	5/23/2012	Active	220	Low terrace, ancient terrace	0.25–0.5	7.75–9.5
Kignèfou	Kérouané	5/23/2012	Active	150	Active channel, low terrace	0.3–0.5	4.5
Sarambali	Kérouané	5/23/2012	Active	300	Low terrace, ancient terrace	0.3–0.6	3.7–4.4
Koboro	Kérouané	5/24/2012	Active	80	Active channel, low terrace	0.25–0.5	3.75–4.5
Foudifimba	Kérouané	5/24/2012	Active	5,500	Ancient terrace	0–18	0
Woulouoro	Kérouané	5/24/2012	Active	600	Active channel, low terrace, ancient terrace	0.15–0.25	5.35–7.75

Deposit Types

Forécariah-Coyah Region

Twelve artisanal diamond mining sites were visited within the Forécariah-Coyah region of western Guinea, from 2010–2012 (table 3; fig. 9). The sites are located in the Atlantic-draining Forécariah River basin. Kourouya 1 and 2, Heremakono, Baschia, and Safoulé are located along tributaries of the Forécariah River; Kenenday, Kenenday North, and Bouramaya are located along a tributary of the Doukourha River; and Banyama, Gbomilo, and Sansangui 1 and 2 are located along tributaries of the Bofon River (figs. 10–14).

Mining within this region takes place within active channel, alluvial flat, and low terrace deposits. The gravel thickness at the visited sites ranges from 0.2 to 2 m, though it is on average 0.5 m thick. Overburden thickness ranges from 0 m in the active channel to 9.5 m in alluvial flat and low terrace deposits, and is on average 2.3 m thick. The underlying bedrock at the majority of the sites (with the exception of Heremanko and Safoulé) is mafic gneiss that is Upper Archean, Lower Proterozoic, or Upper Proterozoic in age. All of the sites, with the exception of Kourouya 1 and 2, were active during the time of fieldwork and had between 3 and 1,200 miners working the deposits.

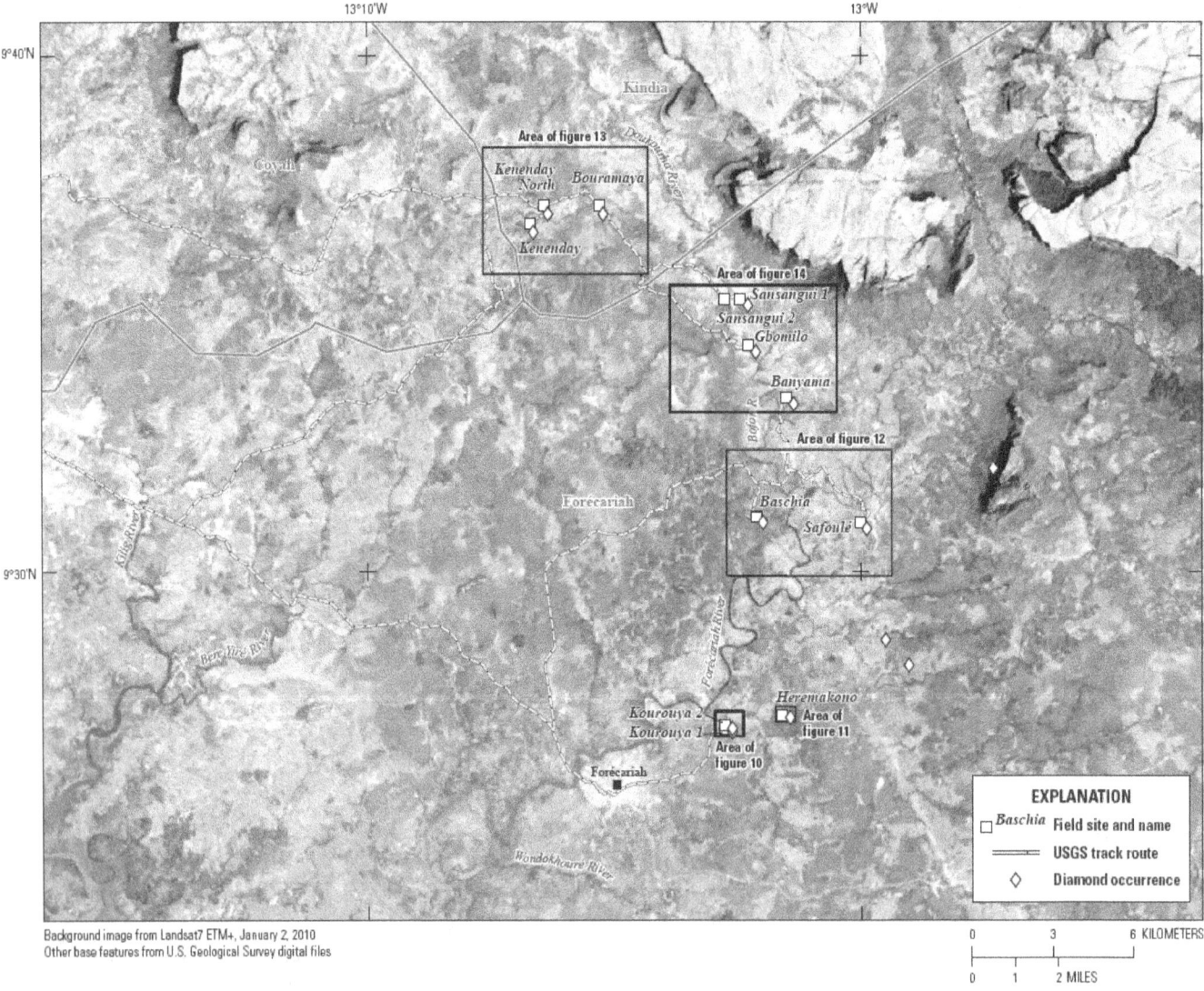

Figure 9.   Site locations within the Forécariah, Coyah, and Kindia Prefectures of Guinea (see fig. 8).

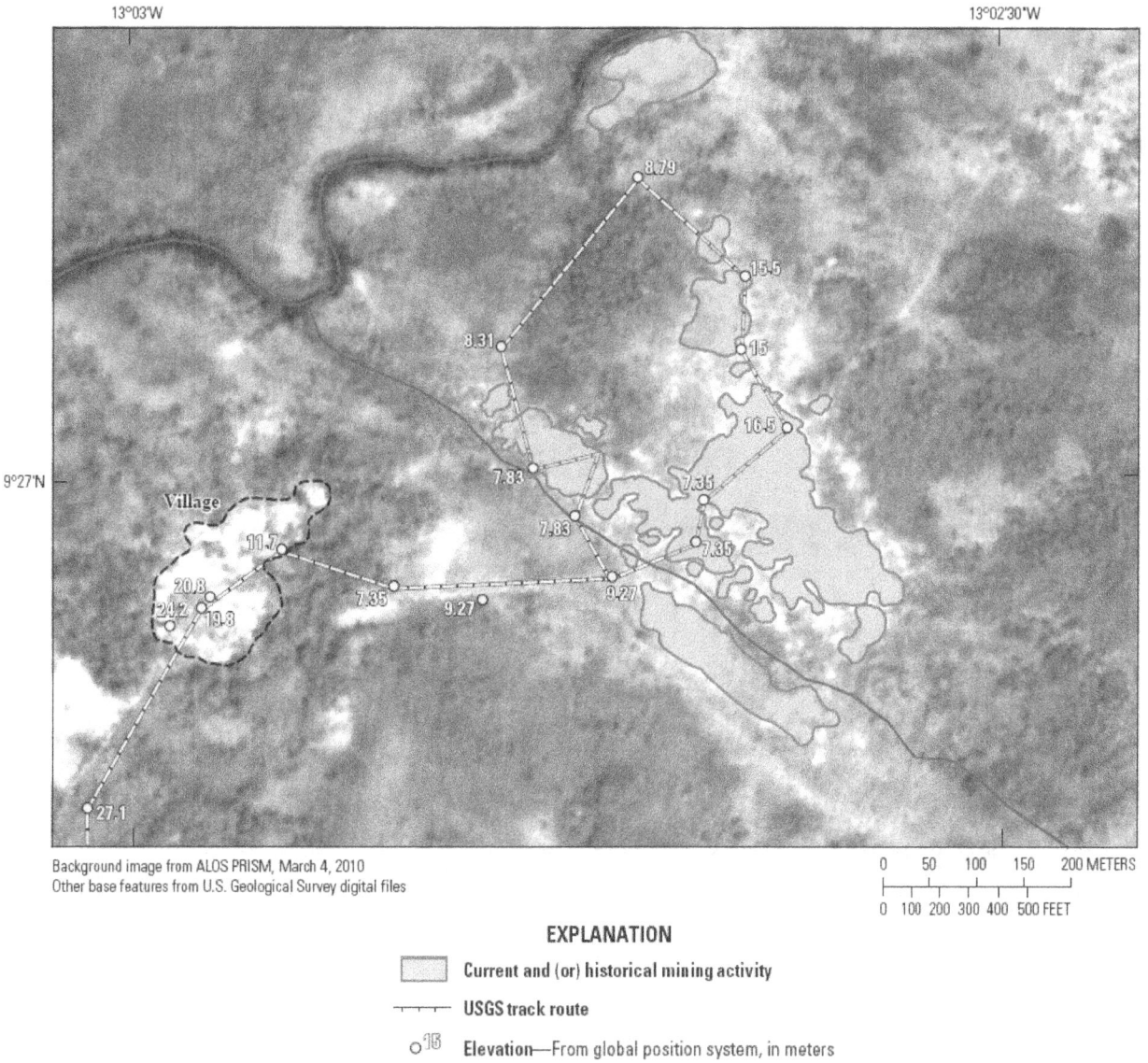

Figure 10.   Satellite image map of Kourouya sites 1 and 2 within the Forécariah Prefecture (see fig. 9).

13°01'30"W

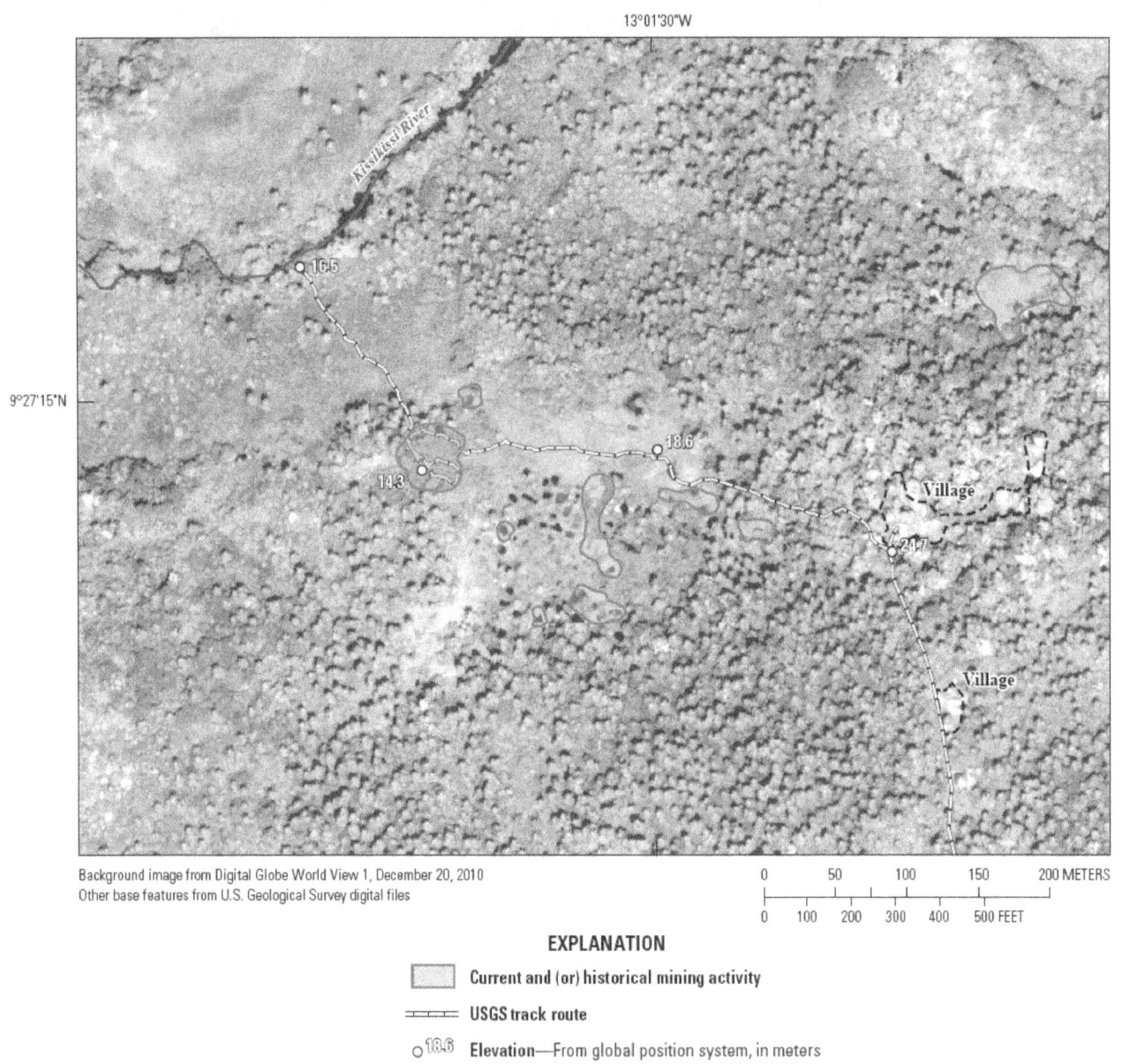

Background image from Digital Globe World View 1, December 20, 2010
Other base features from U.S. Geological Survey digital files

0   50   100   150   200 METERS

0   100   200   300   400   500 FEET

EXPLANATION

Current and (or) historical mining activity

USGS track route

○ 18.6   Elevation—From global position system, in meters

Figure 11.   Satellite image map of the Heremakono field site within the Forécariah Prefecture (see fig. 9).

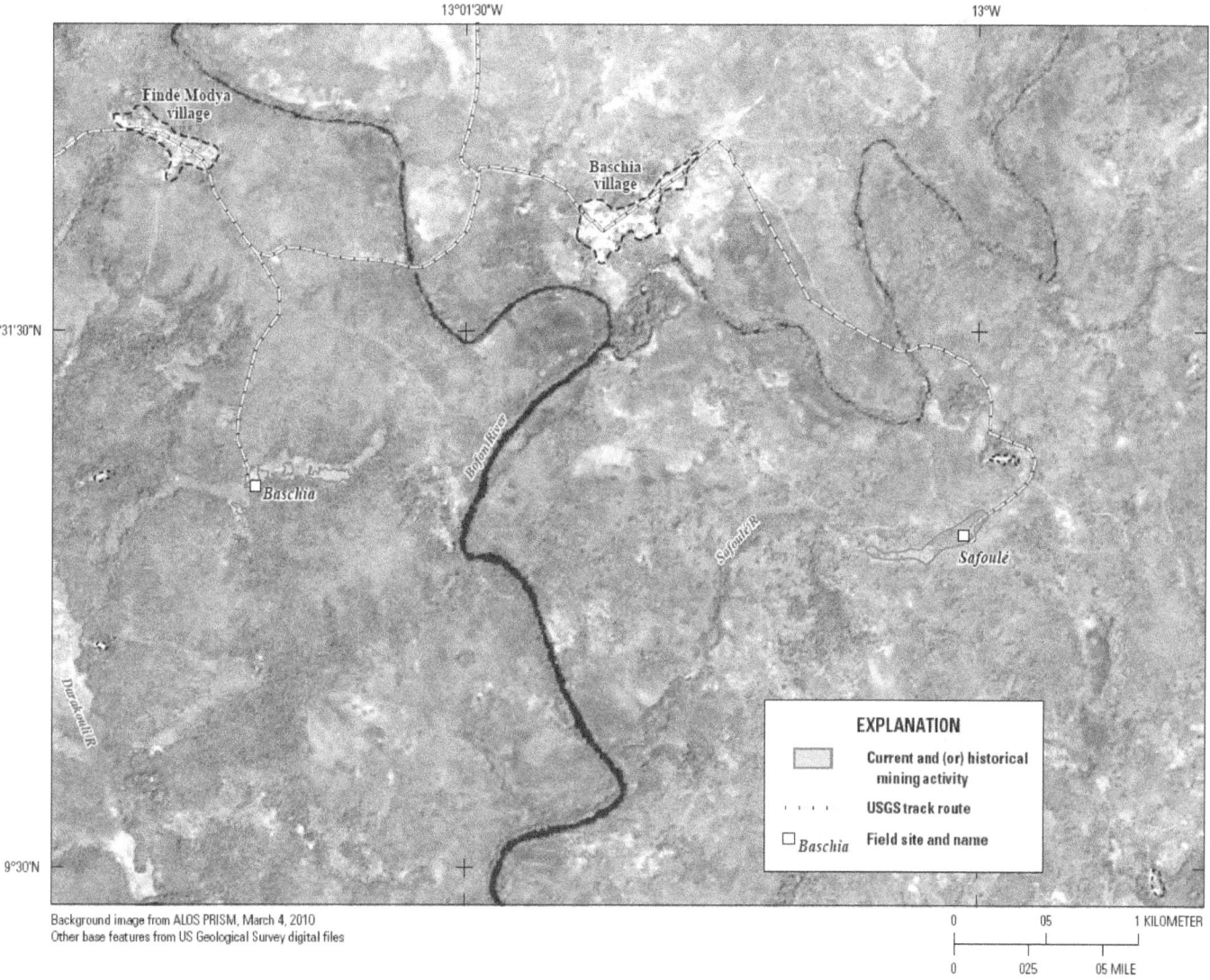

Figure 12.   Satellite image map of the Baschia and Safoulé field sites within the Forécariah Prefecture (see fig. 9).

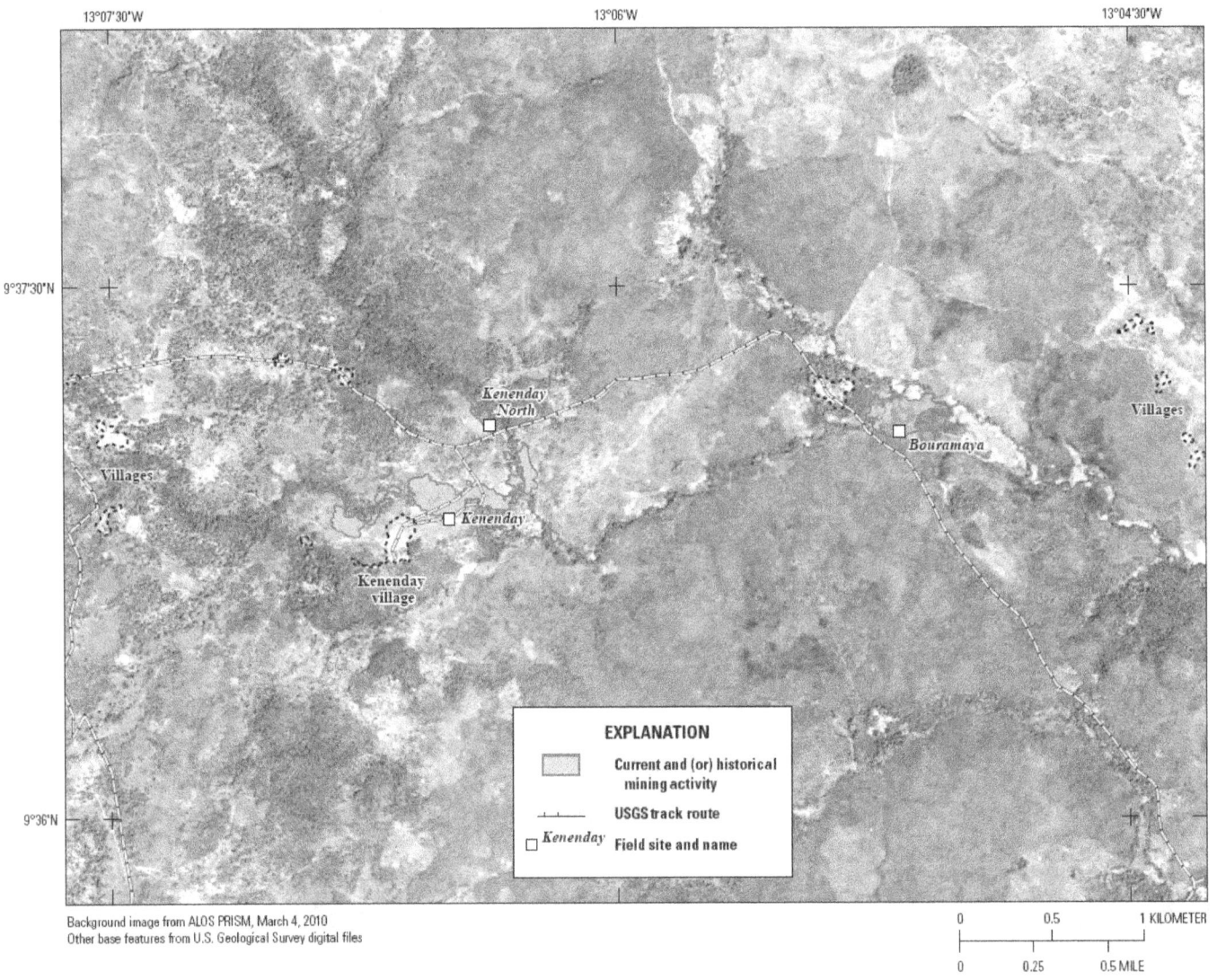

Figure 13. Satellite image map of the Kenenday, Kenenday North, and Bouramaya field sites within the Coyah/Kindia Prefectures (see fig. 9).

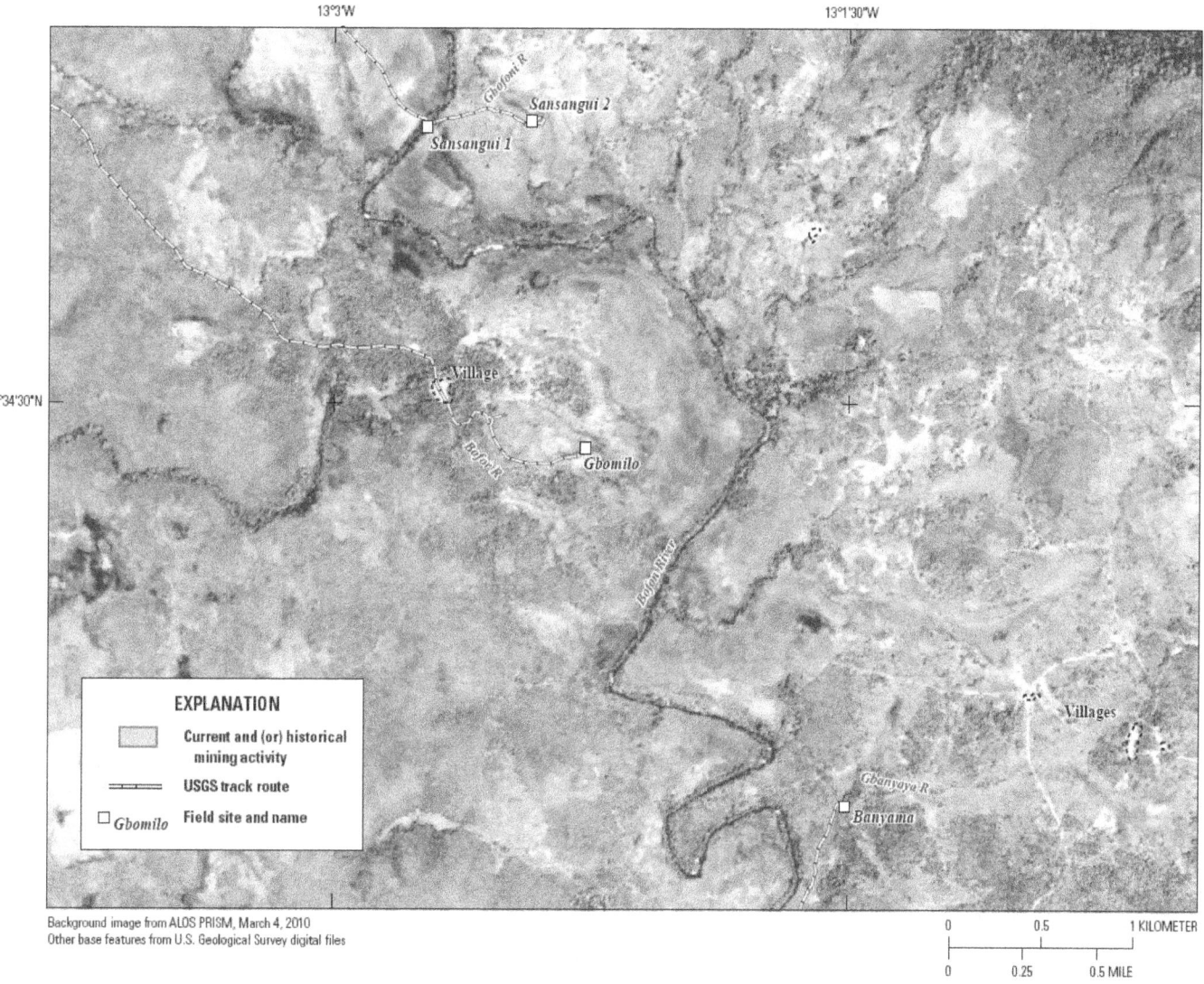

Figure 14.   Satellite image map of Sansangui sites 1 and 2, Gbomilo, and Banyama within the Forécariah Prefecture (see fig. 9).

Kindia-Télimélé Region

Nine artisanal mining sites were visited in the Kindia-Télimélé region of western Guinea in 2010 and 2012 (table 1; fig. 15). The sites are located in the Atlantic-draining Konkouré River basin. The northernmost sites, Kafour, Férékouré, and Témé are located on the Konkouré River, which forms the border between the Kindia and Télimélé Prefectures (fig. 16). The remaining sites are located along tributaries of the Kilissi River (figs. 17–19). At six of the sites, paleochannel deposits filled with sediments and diamondiferous gravels were being mined. At the remaining sites, active channel and low terrace deposits were being mined. The gravel thickness at the visited sites ranges from 0.5 to 1.5 m and averages 1.3 m thick. The overburden layer in this region is relatively thick, ranging from 3 to 15 m and averaging 6.9 m thick. The underlying bedrock at the sites is a Paleozoic conglomerate of sandstone, siltstone, and argillite. All of the sites, with the exception of Foulaya and Samoriya 1, were active during the time of fieldwork and were relatively small operations with the number of miners ranging from 3 to 60.

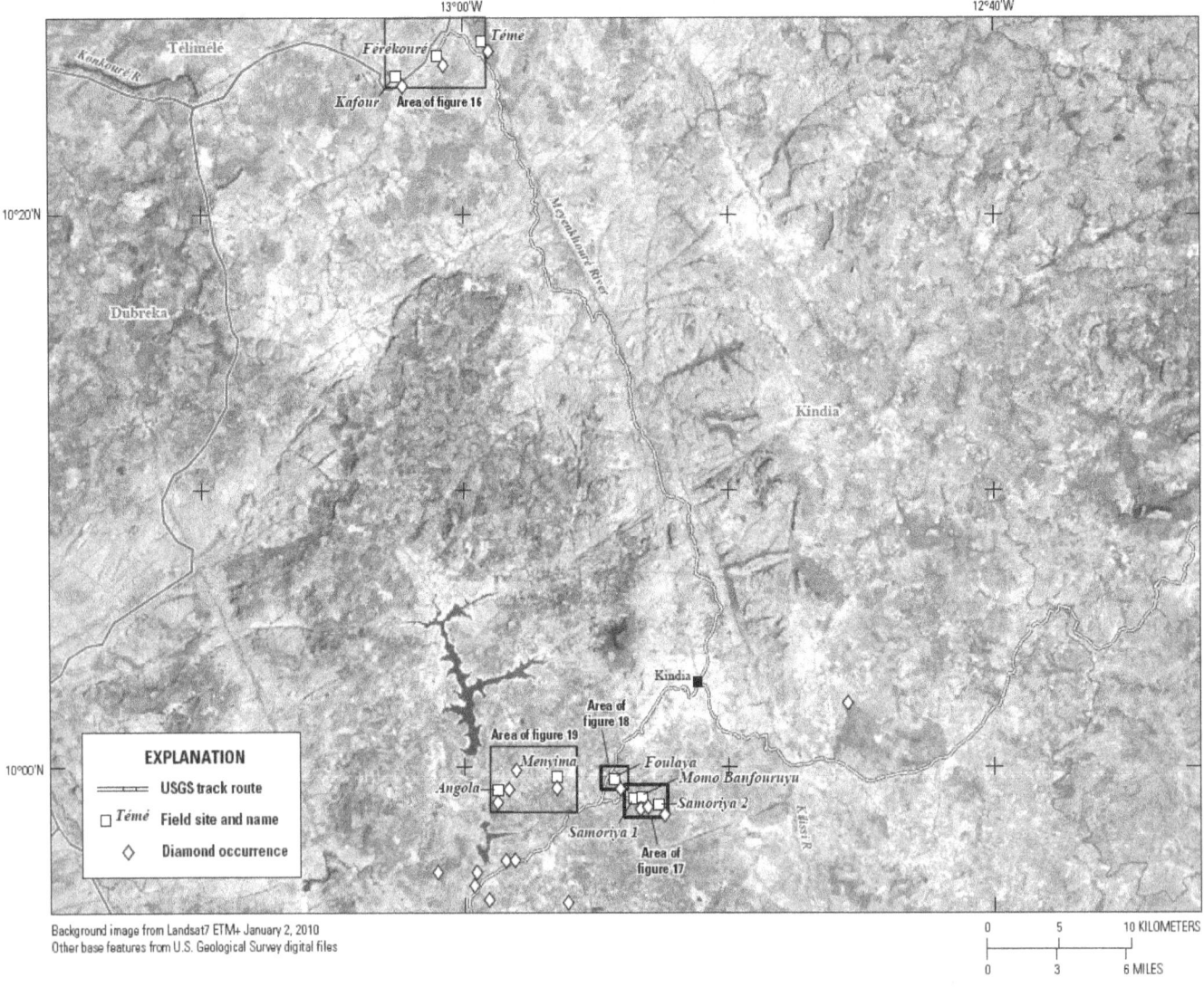

Background image from Landsat7 ETM+ January 2, 2010
Other base features from U.S. Geological Survey digital files

Figure 15.   Site locations within the Kindia-Télimélé Prefectures of Guinea (see fig. 8).

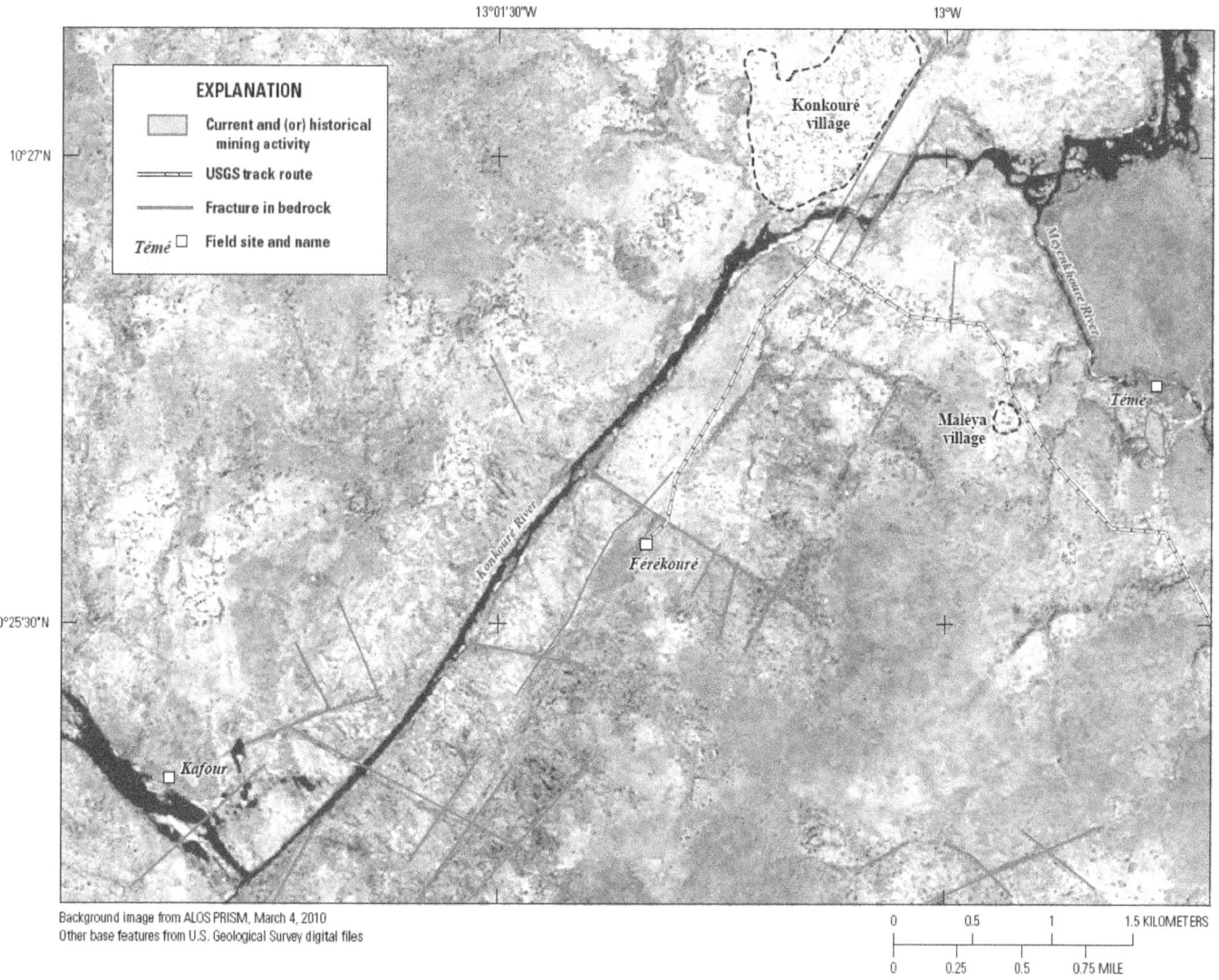

Figure 16.   Satellite image map of the Kafour, Férékouré, and Témé field sites in the Kindia and Télimélé Prefectures (see fig. 15).

12°53'W

9°59'N

Momo
Banfouruyu
village

Momo Banfouruyu □

□
Samoriya 1

Foudjokouré River

Samoriya 2 □

Samoriya
village

EXPLANATION

Current and (or) historical
mining activity

· · · ·   USGS track route

Fracture in bedrock

Samoriya 2 □ Field site and name

Background image from ALOS PRISM, January 17, 2010
Other base features from U.S. Geological Survey digital files

0   250   500 METERS

0   500   1,000 FEET

Figure 17.   Satellite image map of Samoriya sites 1 and 2 and Momo Banfouruyu in the Kindia Prefecture (see fig. 15).

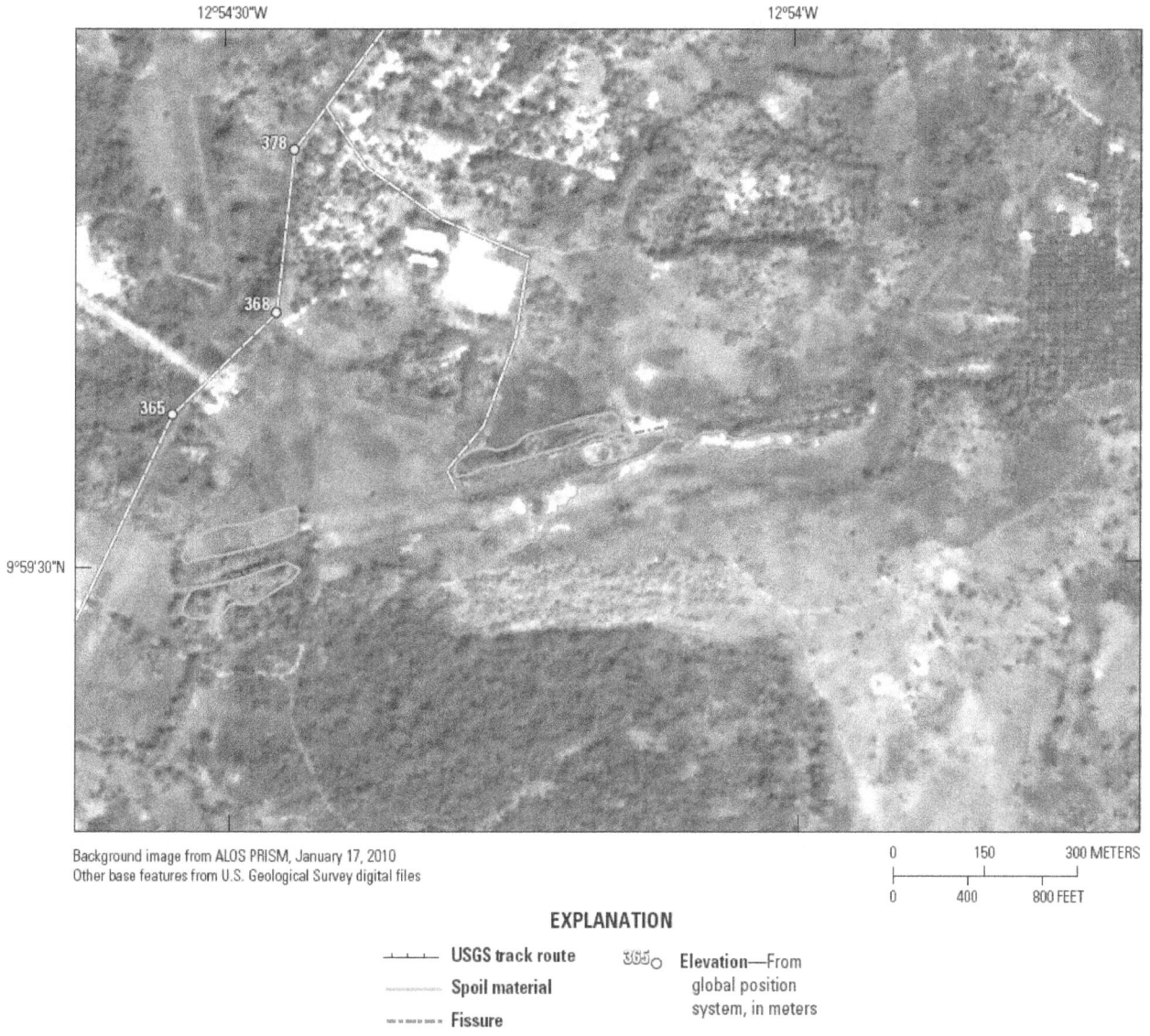

12°54'30"W   12°54'W

378

368

365

9°59'30"N

Background image from ALOS PRISM, January 17, 2010
Other base features from U.S. Geological Survey digital files

0   150   300 METERS

0   400   800 FEET

EXPLANATION

USGS track route

Spoil material

Fissure

365○   Elevation—From
        global position
        system, in meters

Figure 18.   Satellite image map of the Foulaya field site within the Kindia Prefecture (see fig. 15).

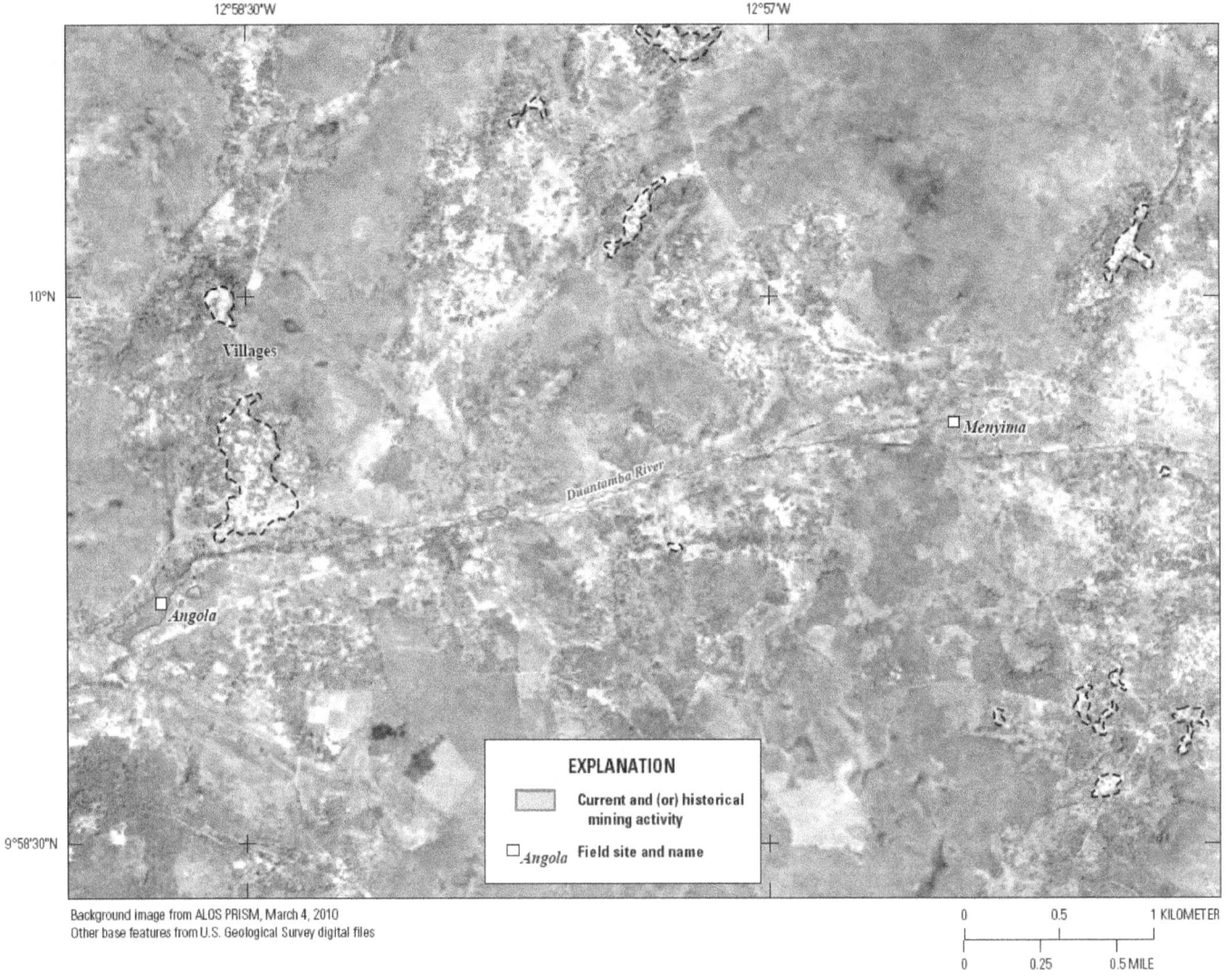

Figure 19.   Satellite image map of the Angola and Menyima field sites within the Kindia Prefecture (see fig. 15).

Kissidougou-Macenta Region

Ten artisanal mining sites were visited in the Kissidougou-Macenta region of southeastern Guinea in 2010 and 2012 (table 1; fig. 20). Six of the sites—Fondiya, Dakodou, Gbanwani, Société Real Rock International (SRRI), Wadagbolofé, and Kogbéla—are located within the Atlantic-draining Moa River basin (figs. 21–25). The remaining four sites are located within the Atlantic-draining St. Paul River basin (figs. 6, 20). The sites in the Moa River basin are located along various tributaries of the Makona River, which forms part of Guinea's southern border with Liberia. The sites within the St. Paul

River basin are located along tributaries of the Diani River. Mining within the Kissidougou-Macenta region takes place within active channel, alluvial flat, and low terrace deposits. The gravel thickness at the visited sites ranges from 0.25 m to 1 m and is on average 0.53 m thick. The overburden layer ranges from 0 m to 6.5 m and is on average 4 m thick. With the exception of two sites, Kolokolo and Bayama, the underlying bedrock at these sites is composed of a Lower Proterozoic undifferentiated complex of migmatic gneiss and porphyritic basalts composed of granites and (or) granodioritics. All of the sites were active during the time of fieldwork and were relatively large, with up to 800 miners working at each site.

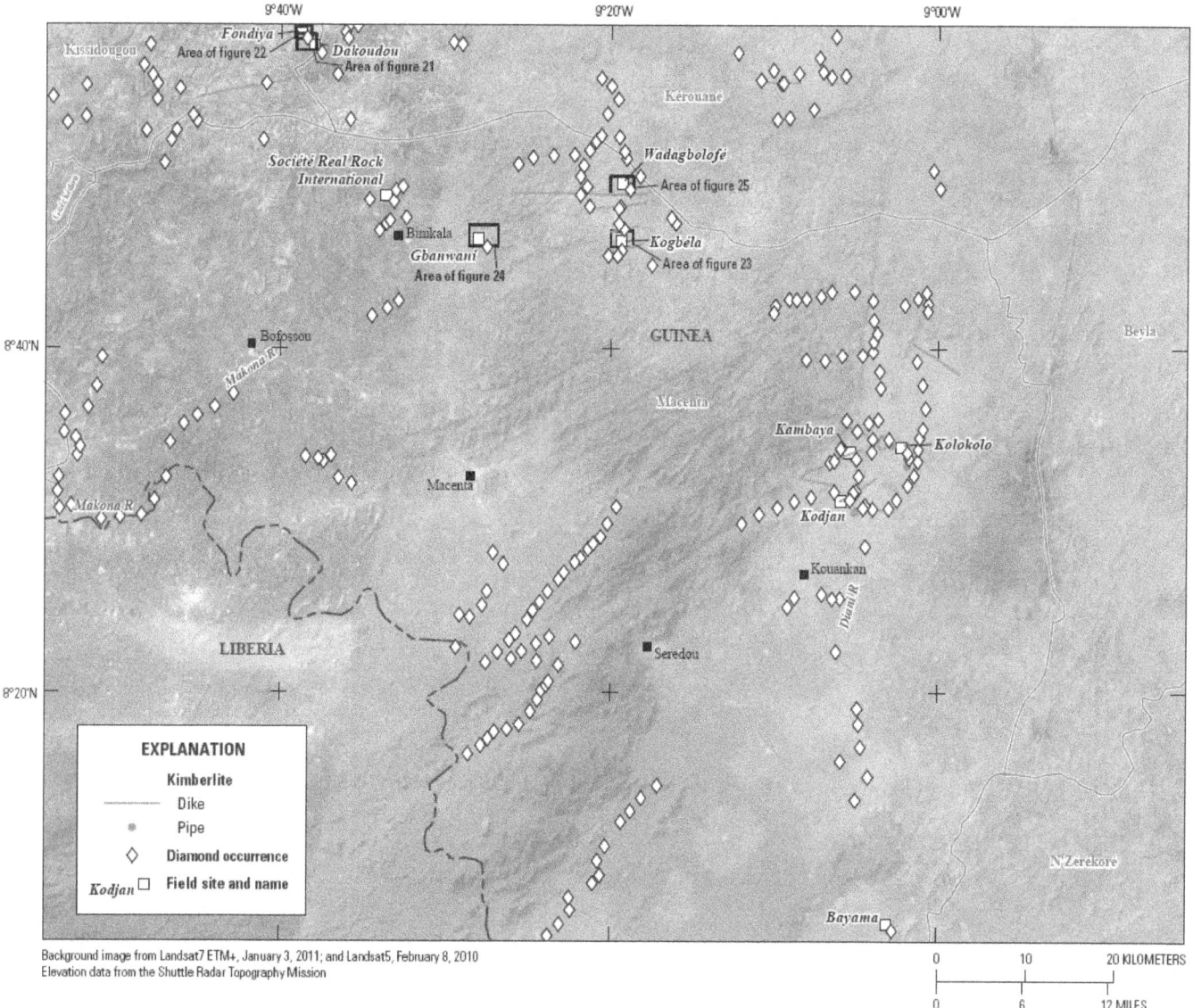

Figure 20.   Site locations within the Kissidougou and Macenta Prefectures of Guinea (see fig. 8).

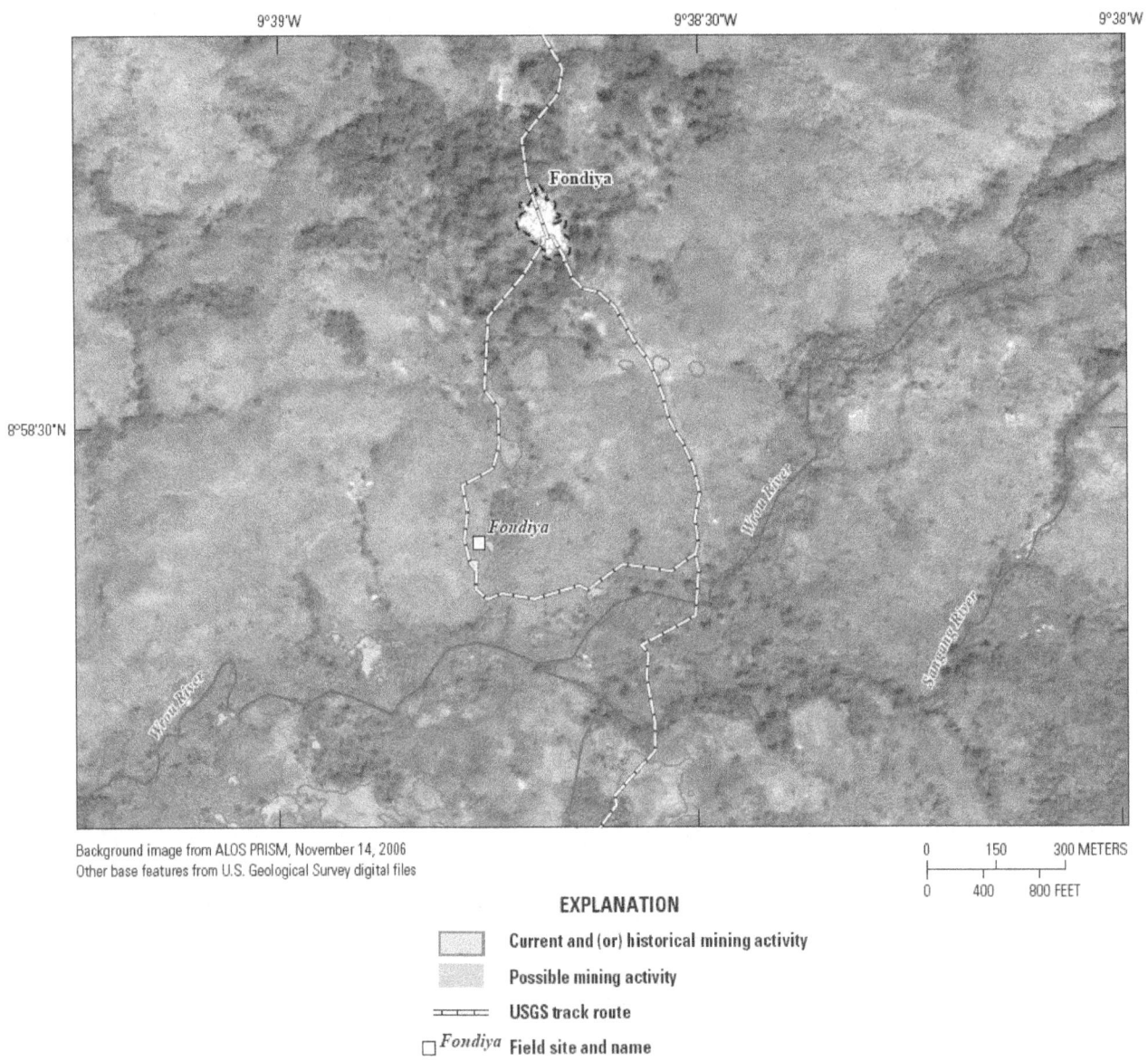

Background image from ALOS PRISM, November 14, 2006
Other base features from U.S. Geological Survey digital files

EXPLANATION

Current and (or) historical mining activity

Possible mining activity

USGS track route

Fondiya Field site and name

Figure 21.   Satellite image map of the Fondiya field site within the Kissidougou Prefecture (see fig. 20).

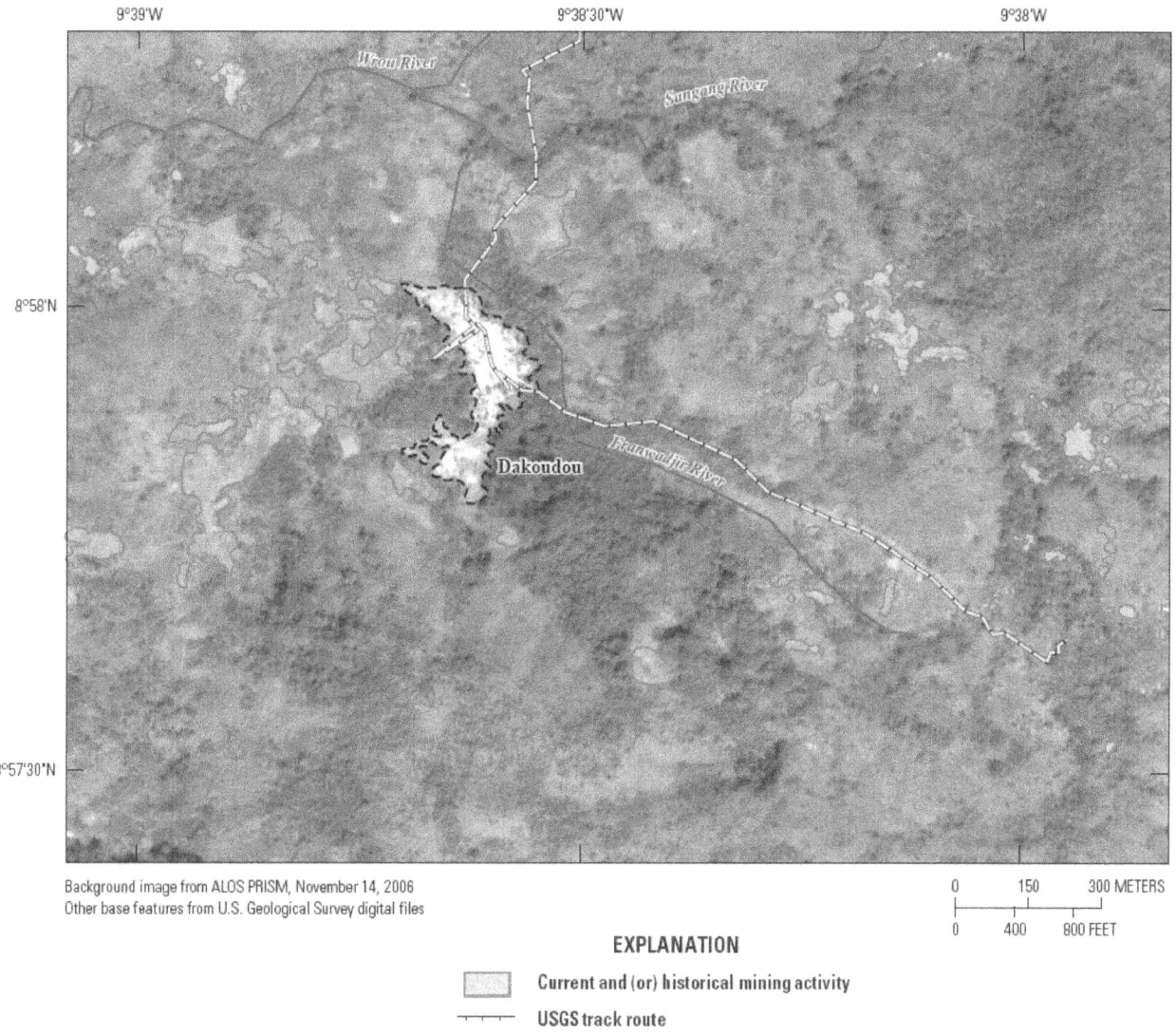

Background image from ALOS PRISM, November 14, 2006
Other base features from U.S. Geological Survey digital files

EXPLANATION

Current and (or) historical mining activity

USGS track route

Figure 22.   Satellite image map of the Dakoudou field site within the Kissidougou Prefecture (see fig. 20).

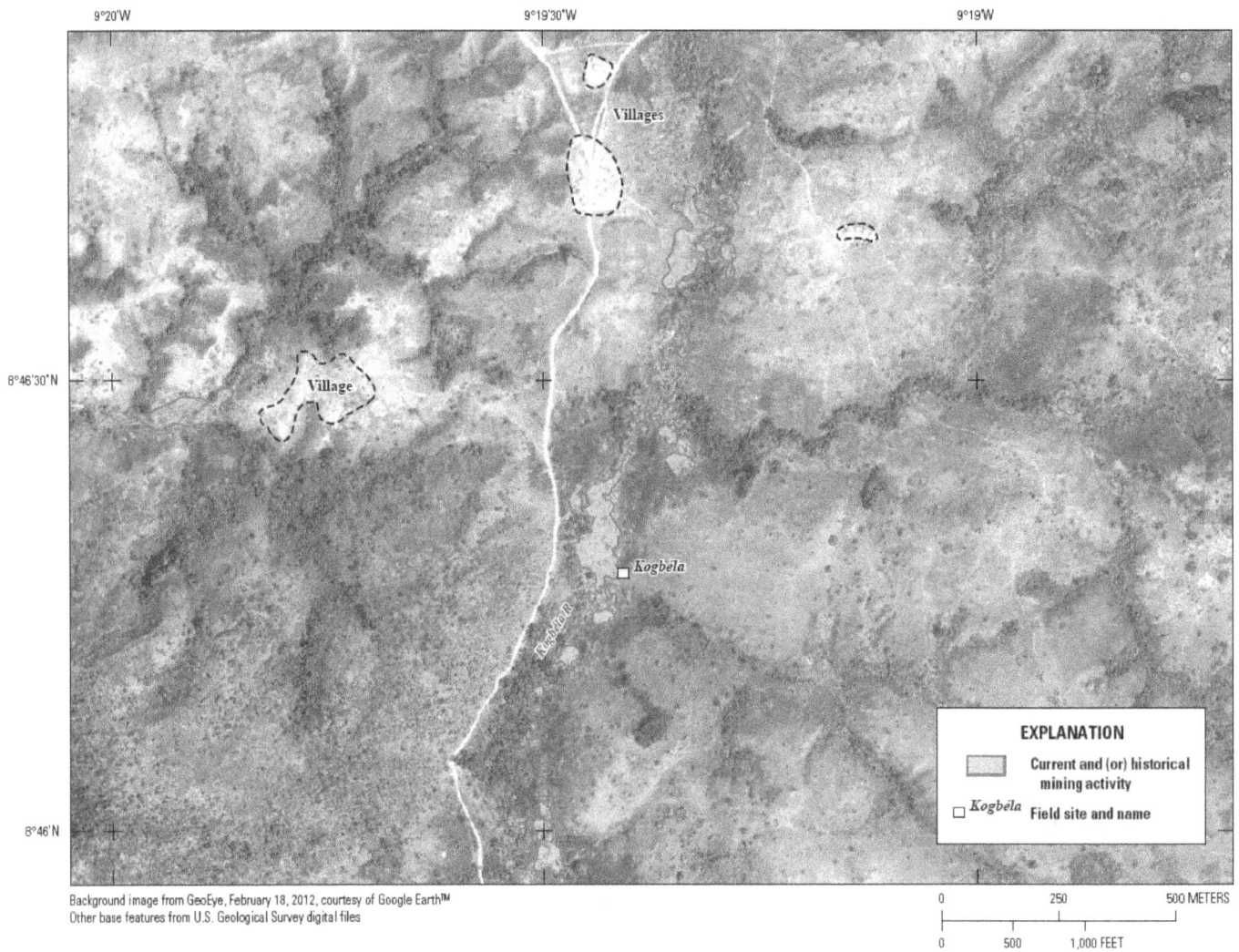

Background image from GeoEye, February 18, 2012, courtesy of Google Earth™
Other base features from U.S. Geological Survey digital files

Figure 23.   Satellite image map of the Kogbéla field site in the Macenta Prefecture (see fig. 20).

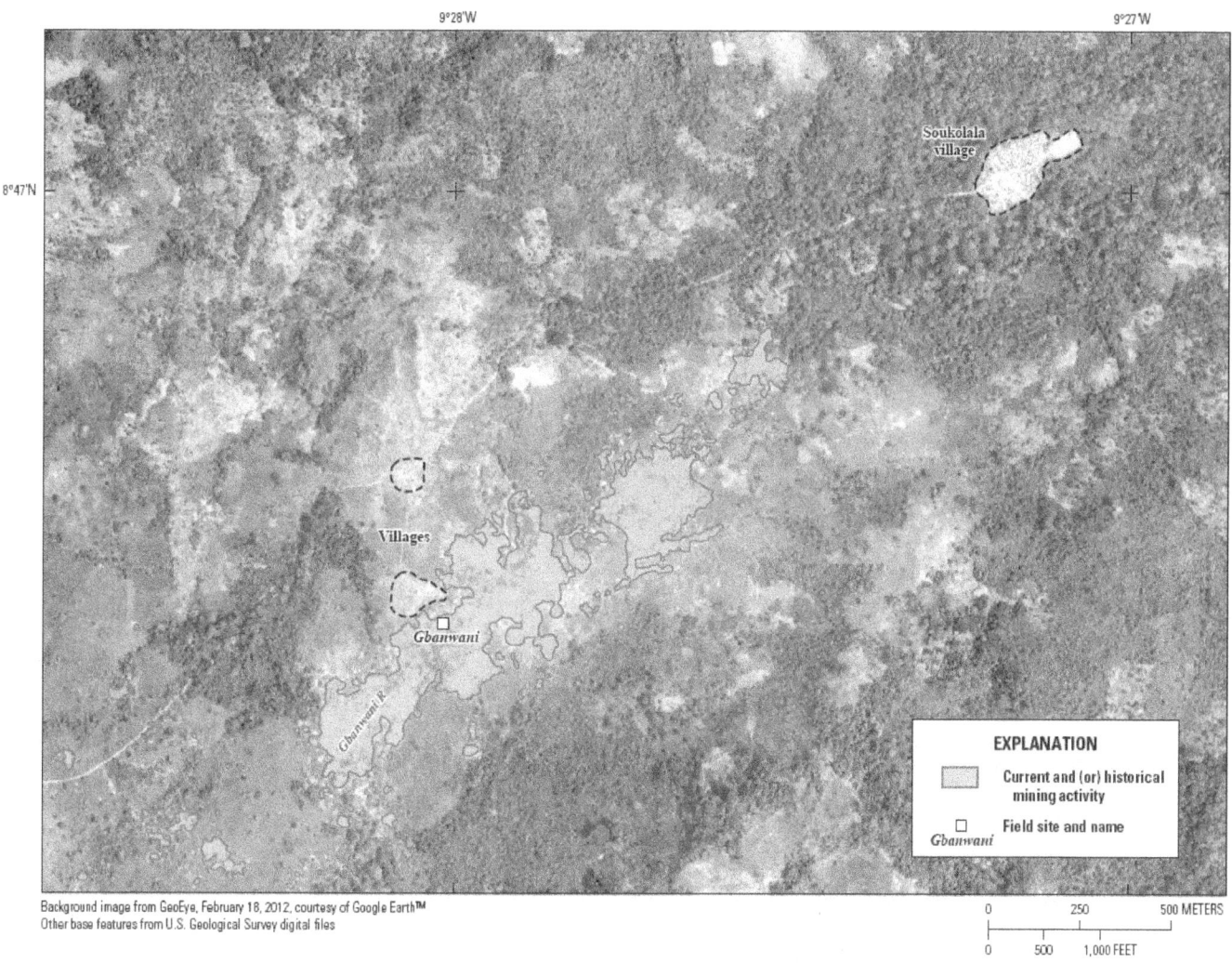

Figure 24. Satellite image map of the Gbanwani field site in the Macenta Prefecture (see fig. 20).

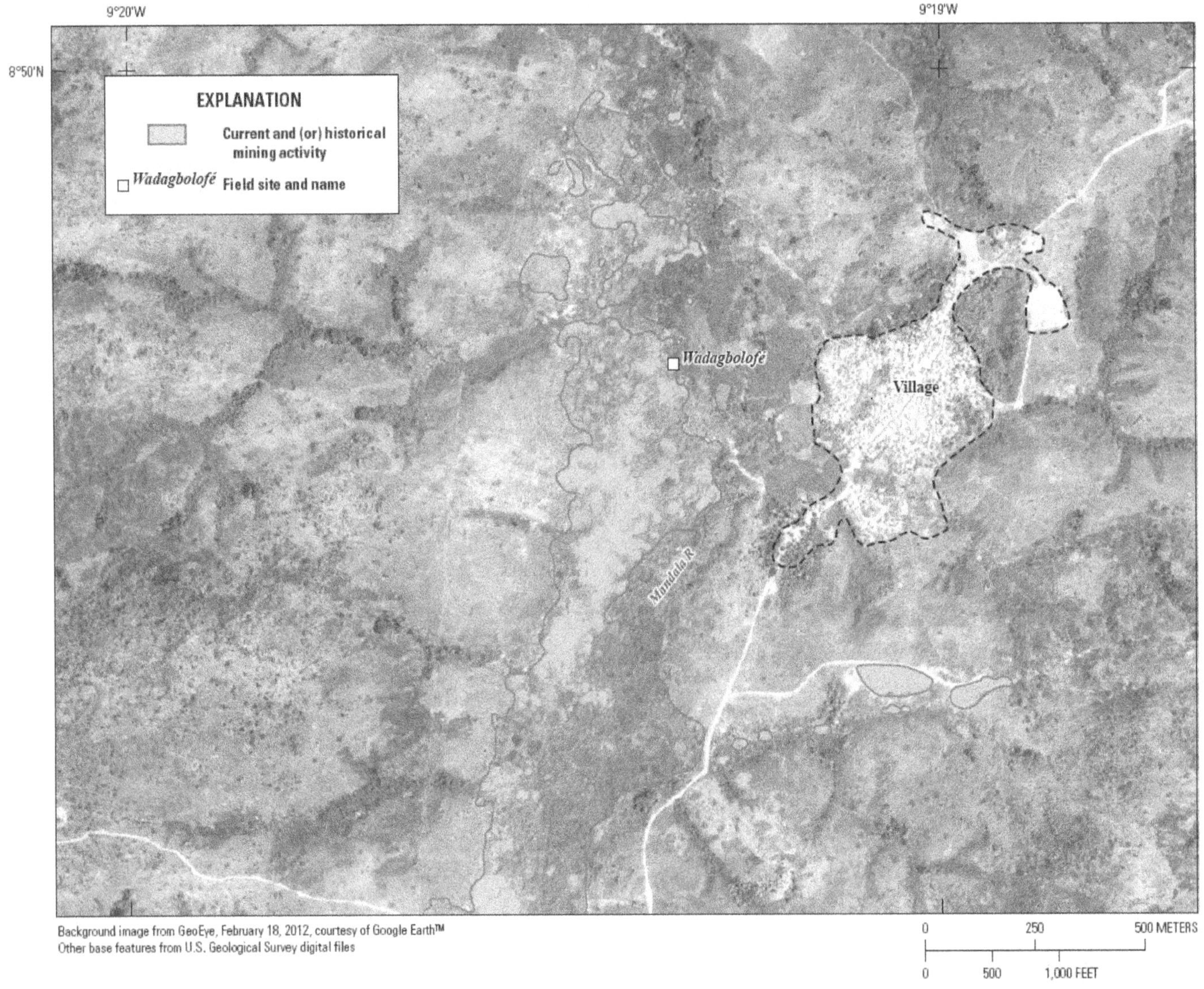

Figure 25.   Satellite image map of the Wadagbolofé field site in the Macenta Prefecture (see fig 20).

Kérouané Region

Nine sites were visited in the Kérouané Prefecture of southeastern Guinea in 2012 (table 1; fig. 26). The Kérouané Prefecture is the location of the Banankoro diamond mining area, which is the most intensively mined region in Guinea. These sites are located within the Niger River basin, along tributaries of the Milo River (figs. 27–31). Mining takes place in active channel, alluvial flat, and low terrace deposits. The gravel thickness of the visited sites ranges from 0 to 18 m thick and is on average 0.4 m thick. Overburden thickness

ranges from 0 to 9.5 m and is on average 5 m thick. Similar to the Kissidougou-Macenta region, the underlying bedrock at the majority of the sites, with the exception of Sarambali, is a Lower Proterozoic undifferentiated complex of migmatic gneiss and porphyritic basalts composed of granites and (or) granodiorites. All of the sites were active during the time of fieldwork. The sites were relatively large, with the number of miners at each site ranging from 20 to 5,500. Table 4 summarizes the soil profile information collected at select sites visited during the field campaign.

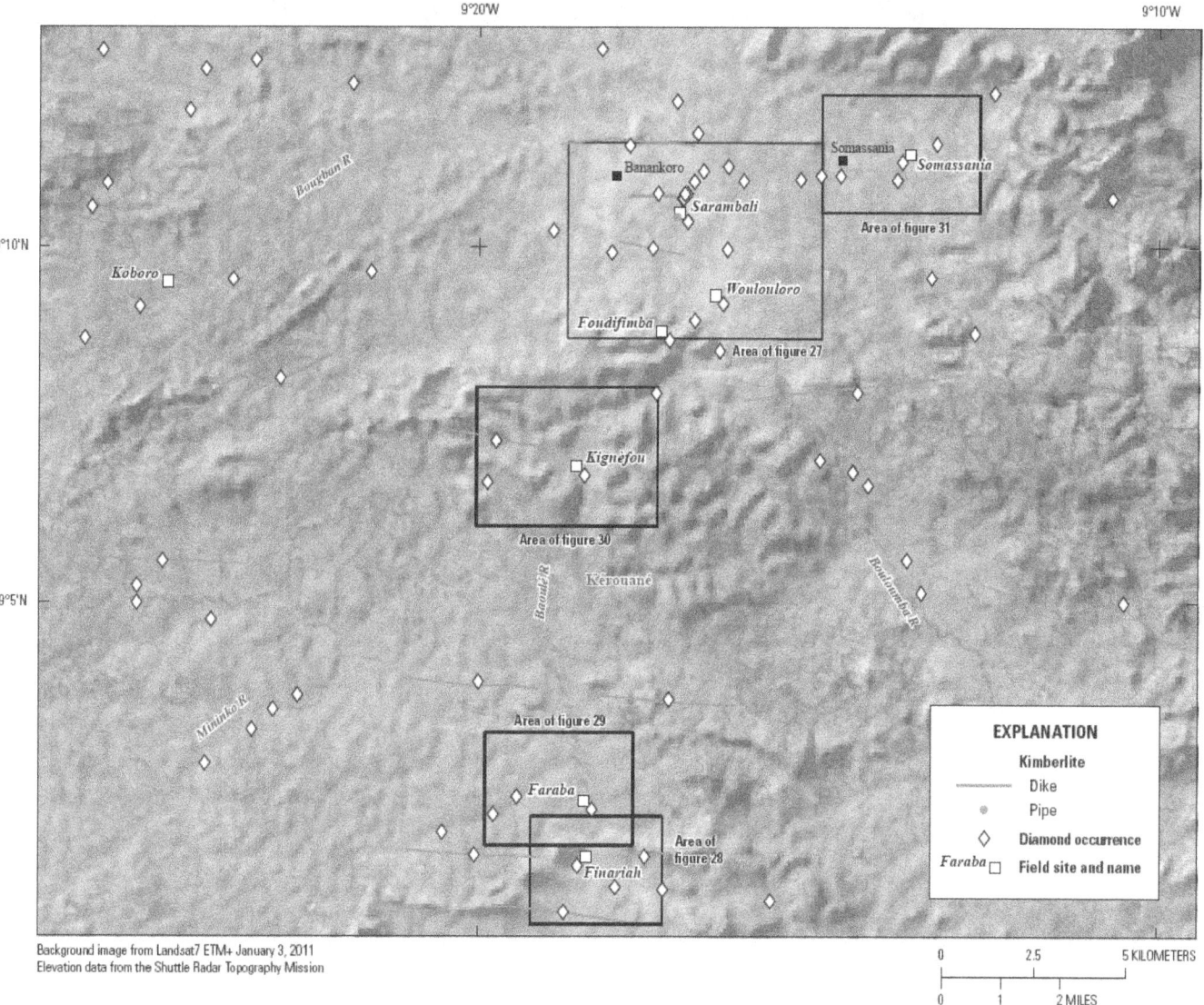

Background image from Landsat7 ETM+ January 3, 2011
Elevation data from the Shuttle Radar Topography Mission

Figure 26.   Site locations within the Kérouané Prefecture of Guinea (see fig. 8).

Background image from the Centre National d'Etudes Spatiales (CNES) Satellite Pour l'Observation de la Terre (SPOT)
(date unknown), courtesy of Google Earth™. Other base features from U.S. Geological Survey digital files

Figure 27.   Satellite image map of the Sarambali, Woulouloro, and Foudifimba field sites within the Kérouané Prefecture (see fig. 26).

Background image from the Centre National d'Etudes Spatiales (CNES) Satellite Pour l'Observation de la Terre (SPOT) (date unknown), courtesy of Google Earth™. Other base features from U.S. Geological Survey digital files

Figure 28.   Satellite image map of the Finariah site in the Kérouané Prefecture (see fig. 26).

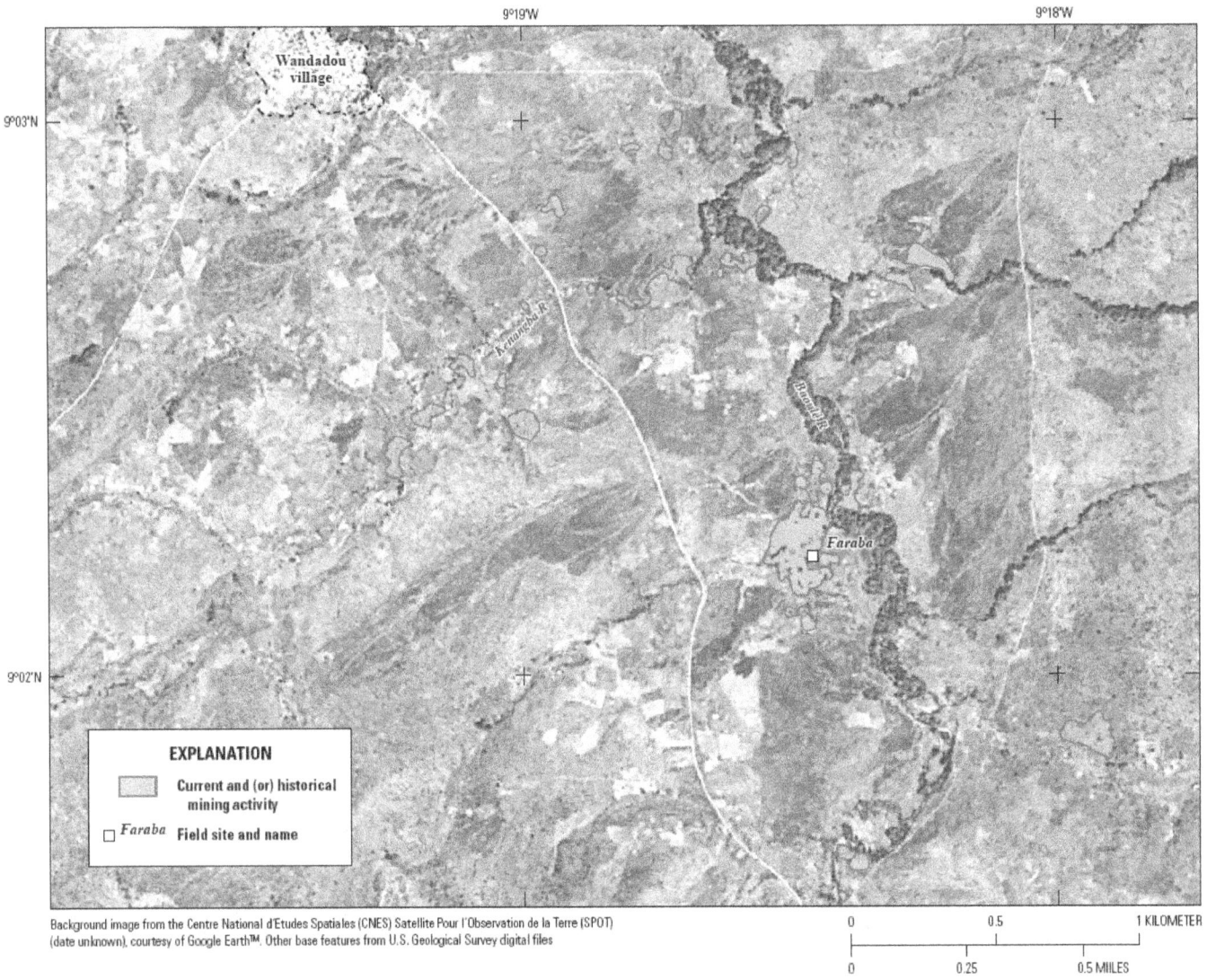

Figure 29.   Satellite image map of the Faraba field site within the Kérouané Prefecture (see fig. 26).

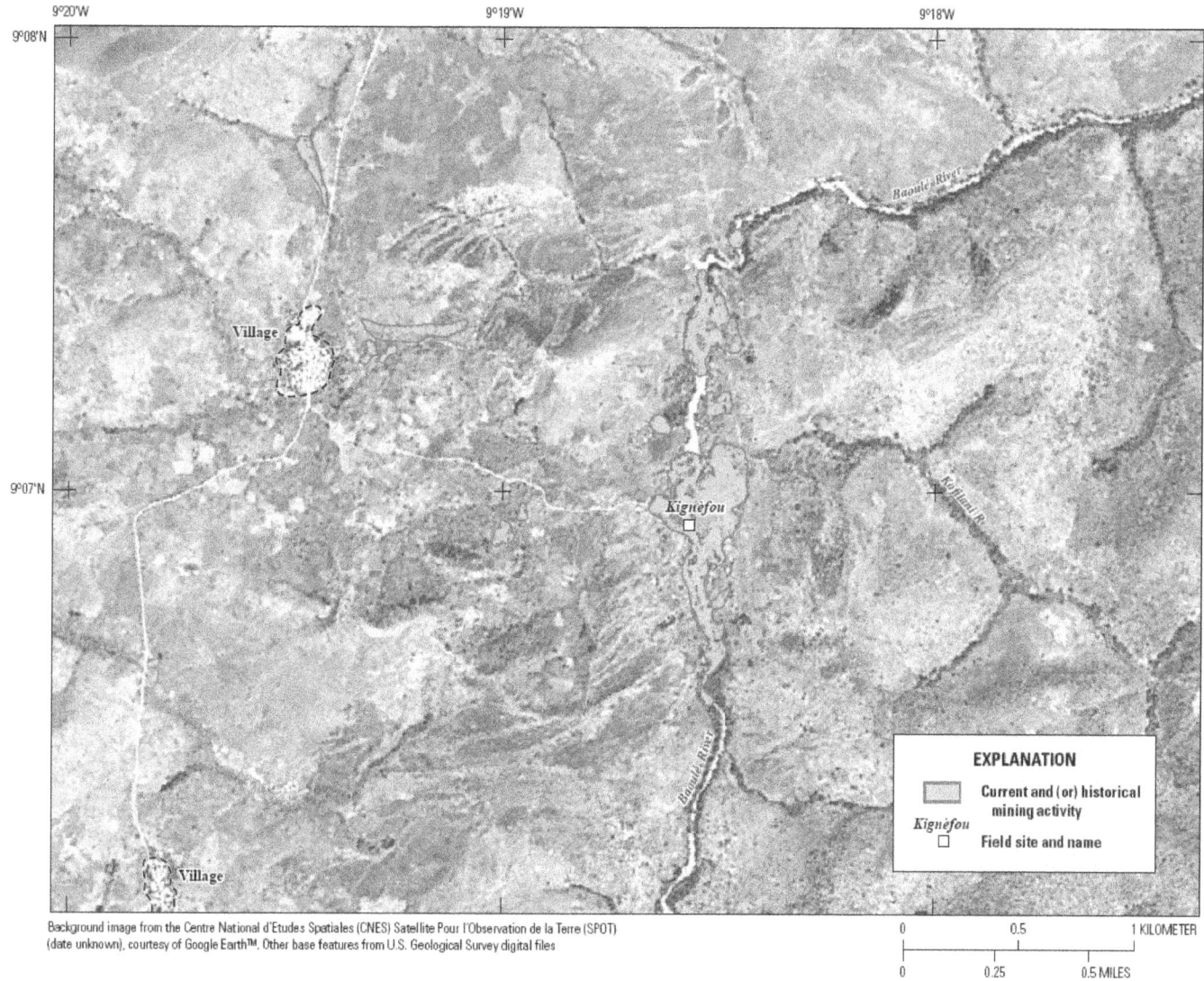

Figure 30.   Satellite image map of the Kignèfou field site within the Kérouané Prefecture (see fig. 26).

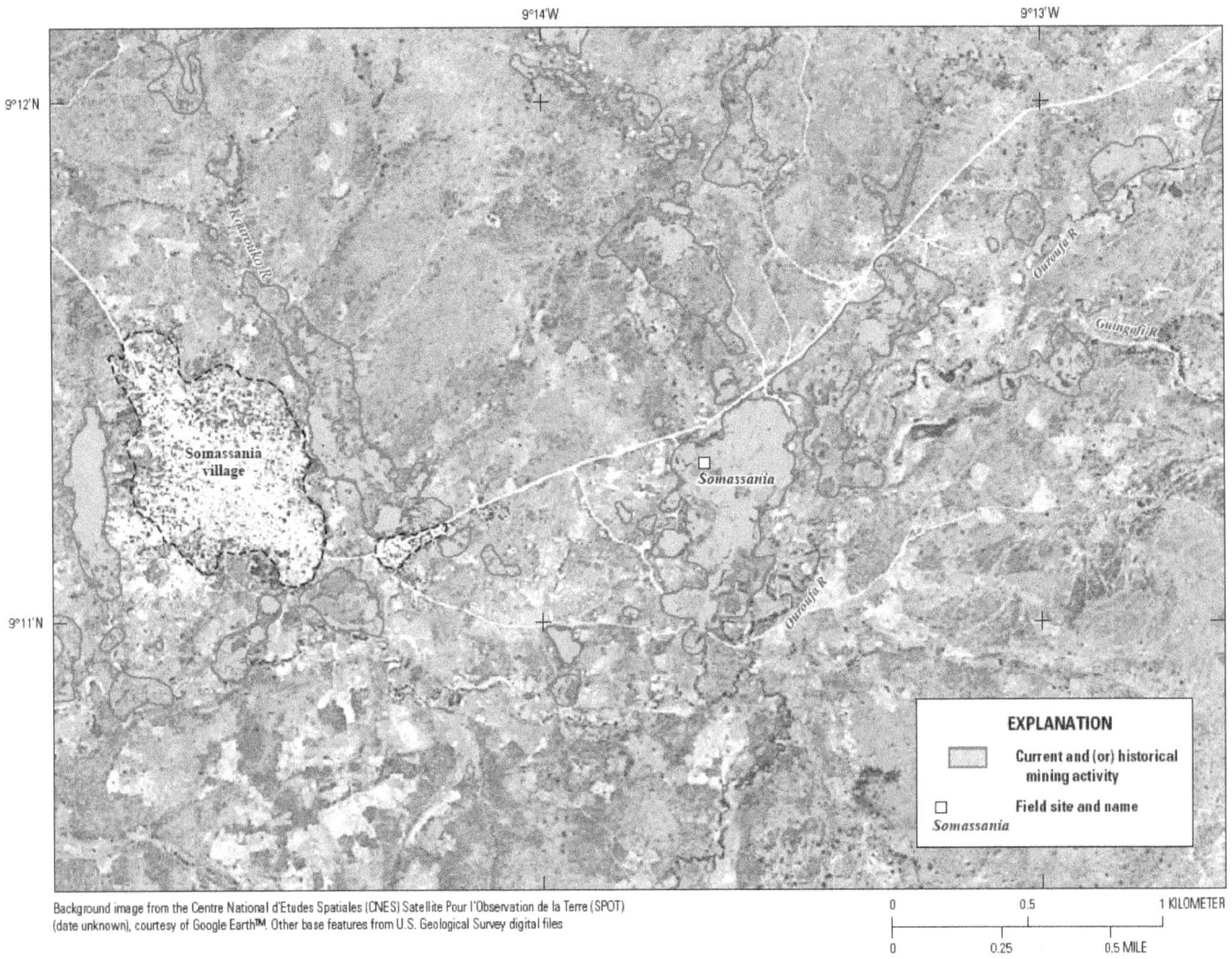

Background image from the Centre National d'Etudes Spatiales (CNES) Satellite Pour l'Observation de la Terre (SPOT) (date unknown), courtesy of Google Earth™. Other base features from U.S. Geological Survey digital files

Figure 31. Satellite image map of the Somassania field site within the Kérouané Prefecture (see fig. 26).

Table 4. Soil profile data obtained during fieldwork in Guinea.

[m, meter; —, unknown]

Site	Prefecture	Easting (m)	Northing (m)	Elevation (m)	Date	River valley	Drainage basin	Geomorphology	Gravel thickness (m)	Overburden thickness (m)
Kourouya Site 1	Forécariah	714589	714589	6.85	4/27/2010	Tributary of Forécariah River	Forécariah	Low terrace	0.4	1
Kourouya Site 2	Forécariah	714590	1045230	8.31	4/27/2010	Tributary of Forécariah River	Forécariah	Alluvial flat	0.5	1.7
Baschia	Forécariah	715688	1052712	23.2	4/27/2010	Tributary of Bofon River	Forécariah	Alluvial flat	0.25	1.5
Heremakono	Forécariah	716679	1045627	14.3	6/3/2011	Kissikissi River	Forécariah	Alluvial flat	0.5	1.5
Banyama	Forécariah	716762	1056977	19.3	3/8/2012	Gbanyaya River	Forécariah	Alluvial flat	0.5	1.5–2
Gbomilo	Forécariah	715356	1058813	26.01	3/7/2012	Bofor River	Forécariah	Active channel	0.3–0.5	1.5–4
Sansangui 1	Forécariah	715059	1060495	21.62	3/7/2012	Bofor River	Forécariah	Low terrace, ancient terrace	0.2–2	1
Safoulé	Forécariah	719515	1052508	33.34	3/8/2012	Safoulé River	Forécariah	Former low terrace	0.5–1	1–2
Kenenday North	Coyah/Kindia	707838	1063750	32.3	3/7/2012	Tributary of Doukourha River	Forécariah	Active channel, low terrace	0.2–0.5	4
Bouramaya	Kindia	7110047	1063785	29	3/6/2012	Tributary of Doukourha River	Forécariah	Active channel, low terrace	0.5	2.5–9.5
Férékouré	Kindia	717104	1153563	183	4/28/2010	Konkouré River	Konkouré	Paleochannel	4	8
Foulaya	Kindia	729648	1105334	385.7	4/30/2010	Tributary of Duaniamba River	Konkouré	Paleochannel	0.5	6
Samoriya 1	Kindia	731057	1104000	—	4/30/2010	Tributary of Duaniamba River	Konkouré	Paleochannel	0.5	5
Fondiya	Kissidougou	429140	992564	676	4/29/2010	Tributary of Wrou River	Moa	Alluvial flat, low terrace	0.5	7.6
Dakoudou	Kissidougou	430468	990570	660	4/29/2010	Franwadji River, a tributary of the Wrou River	Moa	Alluvial flat, low terrace	0.5	3
Kogbéla	Macenta	464422	969607	—	5/22/2012	Kogbéla River	Moa	Active channel, low terrace, ancient terrace	0.25–0.5	3.75–4.5
Finariah	Kérouané	466252	997487	—	5/22/2012	Baoulé River	Niger	Alluvial flat	0.25–0.5	4.75–5.5

Methodology

Diamond Assessment Methods

A historically accepted methodology for assessing diamond deposits involves a mechanized process using costly equipment to sample a site. Before beginning the exploration process, environmental factors such as the regional geology, geomorphology, climate, and the lithologies of the gravel are studied. This information helps to determine which areas should be targeted for exploration, based on the potential for finding economically viable deposits. A detailed regimen entailing shallow drilling, pit excavation, and the concentration of diamonds from gravel is followed. Once a deposit is identified, its economic viability is assessed through a process of reconnaissance pattern drilling and bulk sampling (Marshall and Baxter-Brown, 1995). An effective method of bulk sampling requires creating a grid of pits or parallel trenches based on the initial drilling results (Rombouts, 1995).

While this method is detailed and comprehensive, it is only applicable to sites that are relatively small in area. Following such a method at the regional or country scale is neither practical nor feasible. To assess artisanal mining activity throughout a large region or for the entire country, a different method must be employed. A successful alternative for accomplishing broad-area diamond assessments involves analyzing data collected through field mapping, interviews, and the analysis of remote sensing and geographic information system (GIS) data.

Components of Diamond Resource and Production Capacity Assessments

Bibliographic and Historical Research

The first step of this study is the research, collection, and organization of all available data related to the diamond resources and production in Guinea. All available reports documenting occurrences of diamonds were collected, including reports by private mining companies, geophysical studies, geologic maps, and journal articles. A digital database of diamond occurrences was developed from these sources. Data on overburden thickness, gravel thickness, and grade documented in these reports were used in combination with data collected during fieldwork to model the alluvial deposits.

Field Mapping

Field mapping was conducted at several sites within the study area to develop a database of information on the terrain, geomorphology, geology, mineral deposits, and current mining activities. This methodology took advantage of artisanal mining pits that were already dug, rather than the excavating of new pits. Field mapping is a critical step in the assessment of an alluvial diamond deposit and is composed of several components. First, the geomorphology of the area was mapped. This involves gathering data on the location and elevation of alluvial flats and terraces, using a GPS. Next, individual mine sites were visited, and measurements of the thickness of the overburden as well as a description of the overburden layers from topsoil to bedrock were recorded. The sterile and diamondiferous gravel layers were then measured and described, and the ratio of overburden to gravel depths was recorded. When possible, samples of diamonds recovered at the site were viewed and photographed to provide a qualitative assessment of the characteristics of the stones.

Remote Sensing and GIS

After the conclusion of fieldwork, satellite images of the study sites were analyzed, and all currently exploited diamond mining areas visible in the imagery, including the mines visited during fieldwork, were mapped. Imagery from the Advanced Land Observing Satellite (ALOS) PRISM, DigitalGlobe's WorldView-2, and Landsat sensors were used to conduct this analysis (table 5).

Launched by the Japanese Aerospace Exploration Agency (JAXA) on January 24, 2006, ALOS contains three sensors: AVNIR2, PALSAR, and PRISM. PRISM is a 2.5-m resolution panchromatic stereo imager. Five PRISM scenes covering parts of Forécariah, Kindia, Télimélé, and Kissidougou were acquired from the Remote Sensing Technology Center (RESTEC). The images were imported from their native L1B format and reprojected into the proper Universal Transverse Mercator (UTM) zone. Data collected in the field on the location of active diamond mining sites were used as control points during the process of delineating additional active mining zones visible in the PRISM imagery. These sites were added to the database as either confirmed mining sites or potential mining sites.

Additional imagery was acquired from DigitalGlobe's WorldView-2 sensor. This sensor, launched on October 8, 2008, provides imagery that is 0.5 m resolution in the panchromatic and approximately 2 m resolution in the red, green, blue, and near-infrared multispectral bands. A WorldView-2 image with coverage of the Forécariah Prefecture was obtained for this study.

Additionally, 30-m resolution Landsat7 ETM+ and Landsat5 data scenes covering the Forécariah, Coyah, Kindia, Télimélé, Kissidougou, Macenta, and Kérouané Prefectures were downloaded from USGS's EarthExplorer portal. These data were acquired to provide a consistent imagery base coverage for all of the visited prefectures.

The GIS consists of a database of all known diamond occurrences in Guinea. This database was compiled from existing bibliographic sources of information on Guinean diamond deposits as well as new records from mining company reports, the field mapping of sites during the fieldwork phase of the project, and satellite image interpretation of

Table 5. Satellite imagery scene information used in this study.

[GMT, Greenwich Mean Time; ALOS, Advanced Land Observing Satellite; ETM, Enhanced Thematic Mapper]

Sensor	Prefecture	Center latitude	Center longitude	Image date	Image time (GMT)
ALOS PRISM	Forécariah	9.491	−13.152	March 4, 2010	11:28:47.33
ALOS PRISM	Kindia 1	9.989	−13.045	March 4, 2010	11:28:39.10
ALOS PRISM	Kindia 2	9.935	−12.785	January 17, 2010	11:28:52.70
ALOS PRISM	Kissidougou	8.939	−9.766	November 14, 2006	11:13:55.072
ALOS PRISM	Télimélé/Kindia	10.487	−12.937	March 4, 2010	11:28:30.88
DigitalGlobe WorldView-2	Forécariah	9.491	−13.065	December 20, 2010	11:32:28
Landsat 7 ETM+	Forécariah/Coyah/Kindia/Télimélé	10.129	−13.031	January 2, 2010	11:00:56.82
Landsat 7 ETM+	Kérouané/Macenta	8.733	−8.65	January 3, 2011	10:41:53.96
Landsat 5	Kissidougou	8.803	−9.993	February 8, 2010	10:49:03

mining areas. Each occurrence in the database was attributed with geographic coordinates, date and source information, and data on the geology and geomorphology of the deposit, which includes thickness of overburden, gravel thickness, and grade information, where available.

Hydrological Analysis and Geomorphic Modeling

A number of physical environmental factors are required in order to better understand the location and depositional environment of diamond deposits. Marshall and Baxter-Brown (1995) provide a comprehensive review of the factors that are incorporated in an exploration program. These include regional geology, geomorphology, depositional climate, and postdepositional changes that affect reworking or further concentration of deposits. This study is primarily concerned with alluvial deposits; thus, it is focused on the fluvial geomorphology of the known diamond occurrences. This study uses a digital elevation model (DEM) of Guinea and the surrounding region to model the hydrologic and fluvial geomorphic characteristics of the diamond occurrences. Statistical data on the volume of diamondiferous gravels of the deposits were then developed on the basis of the output of the geomorphic model, data from fieldwork, and data on grade and gravel thickness.

Rivers and streams were ordered from 1 to 8 using the Strahler (1964) method of stream numbering. The logic behind the stream-ordering system developed by Strahler (1964) is that the order number is directly proportional to the size of the contributing watershed, to channel dimensions, and to stream-discharge measurements for each stream segment. These attributes are then used to model the fluvial geomorphology of each of the subbasins within the watersheds.

Previous research has shown that diamond deposit grades are correlated with their location in either terraces or alluvial flats (Rombouts, 1987a; Sutherland and Robinson, 1996). For this reason, modeling of the fluvial geomorphology helps show the geographic extent of alluvial flat and alluvial terrace

landforms for each watershed basin. These landform units can then be measured to obtain an estimate of the volume of diamondiferous gravel remaining in each diamondiferous watershed. The modeling uses a relative relief model of the terrain above the base flow of the closest river segment (Chirico and others, 2010c). This method calculates alluvial flats from base elevation along the river flowpath to not more than 5 m above base elevation. Alluvial terraces are calculated from 5 to 10 m above base elevation. The values used in modeling the alluvial flat and terrace deposits of Guinea are average values derived from the database and from the field mapping of deposits.

All order 1 watersheds which have a record of a diamond occurrence were selected and used to process potentially diamondiferous watersheds. The volume and grade of deposits in the selected watersheds were then calculated on the basis of the surface area of the landform model for alluvial flats and terraces and the available grade and volume data. This analysis was performed for five identified "regions" in Guinea: Forécariah-Coyah, Kindia-Télimélé, Kissidougou-Macenta, Kérouané, and "Other," which is composed of all other watersheds with an occurrence that falls outside the four main regions.

Evaluating the grade of alluvial deposits in Guinea was challenging. The clustering and concentration of diamonds in low-grade but high-value deposits has likely occurred from the successive rejuvenation of rivers and streams and formation of many paleochannels (Ellis, 1987). Rombouts (1987b) addressed the problems in estimating the grade of these types of deposits in Guinea. Sichel's (1973) method, based on a modified Poisson distribution, was found to be the best method for evaluating the AREDOR concession's alluvial deposits. However, in that case, detailed sampling at up to 100 by 50 m grids had been completed and was available for modeling and estimation. Similar data are nonexistent for the deposits in most of the hundreds or thousands of areas currently being mined by artisans throughout Guinea.

There are very few references to grades of alluvial deposits in Guinea. Only 11 records in the database contain reliable grade information. Therefore average grades were derived from the database records based on geomorphological similarities within the watershed zones (fig. 8).

Diamond Resource Potential Methodology

The diamond resource potential was calculated for each of the five regions studied in this assessment. A modified volume and grade approach, developed by Barthélémy and others (2006), was used to calculate the resource potential for each of the five regions, the results of which were totaled to arrive at a total resource potential estimate for Guinea. The volume and grade approach can be described mathematically as:

$$P = (3 / 4V \times T1) + (1 / 4V \times T2), \qquad (1)$$

where P is the estimated diamond resource potential, and V is the volume of alluvium. The volume of alluvium is determined by estimating the width of the alluvial-flat deposit and multiplying it by the thickness of the gravel layer being mined. The product of this multiplication yields a number equal to the volume of diamondiferous gravel. Alluvial-flat widths are determined by the order of the river or stream.

The gravel layers are not all equally endowed with the same content of diamonds. Some gravel deposits within the alluvial flats may have higher concentrations than others, on the basis of depositional history, the type of fluvial environment present, and the time period during which the alluvial gravels were originally deposited. To account for these variations in depositional history, two gravel grades are used in the formula. One grade is described as the "basic" grade and is applied to three-fourths of the alluvial gravels calculated for the volume. The second grade is the "concentration" grade. This value is applied only to one-fourth of the total calculated volume of alluvial gravel in the deposit. $T1$ corresponds to the "basic" grade and $T2$ corresponds to the "concentration" grade of the alluvial gravels (Chirico and others, 2010b).

This analysis was performed on all order 1 watersheds within each of the five regions, based on the geomorphic model's characterization of alluvial flats and terraces, average gravel thicknesses calculated from field data, and average grades calculated from historical records.

Diamond-Production Capacity Methodology

The second component of this study was to determine the diamond-production capacity. Diamond-production capacity is defined as the current volume of diamonds (calculated as total number of carats) which can be produced utilizing current human and physical resources. It is a measure of the current state of the diamond mining sector based on recent field data and on previous research studies of diamond mining, gravel grades, records of mining companies, and estimates of the total number of alluvial diamond diggers and small mining cooperatives. Production capacity was calculated for each of the five regions, the results of which were then totaled, to arrive at a final production capacity estimate for Guinea. The formula used to calculate the production capacity can be described mathematically as:

$$P_i = (Vm / d \times g) \times d \times A_i + Ip, \qquad (2)$$

where P_i is the total current production capacity; Vm/d is the volume of gravel worked per digger per day; g is the average gravel grade; d is the total number of days that a digger works per year; and Ip is the estimated industrial production. Finally, A_i is the total number of diggers actively mining diamonds.

Historical data were used to estimate g and Ip, for each region, while a combination of historical data and fieldwork observations was used to calculate Vm/d, d, and A_i. To calculate Vm/d, data on overburden thickness, gravel thickness, the surface area of pits, the number of days required to exploit a pit, and the number of miners working each pit were used. The necessary data were available for seven field sites. To perform the calculation, the average surface area of the pits at each site was multiplied by the total thickness of the overburden and gravel layers to arrive at the volume of material in each pit. This number was then divided by the number of days required to extract and wash the material, the result of which was divided by the number of miners working each pit, to arrive at the total volume of material dug per person per day. Lastly, the ratio of gravel to total material was calculated and multiplied by the volume of material dug per person per day to arrive at the total volume of gravel dug per person per day (Vm/d). These calculations were performed for seven field sites and were then averaged, resulting in a final value for Vm/d (table 6).

To calculate the number of days miners work per year (d), data collected during fieldwork on which days and months miners work were averaged for each region to estimate the number of days miners work per year at the regional scale.

Two methodologies were used to calculate A_i at the regional and national scales, both relying on (1) the digital database of diamond occurrences, compiled from bibliographic sources, fieldwork data, and satellite image interpretation and (2) data collected in the field on the number of active miners at each site. The database occurrences were used to estimate the number of active diamond mining sites in Guinea, while the field sites were used as a representative sample of active diamond mining sites in Guinea.

The first of these methods is an Averaging Method. To calculate the average number of miners using this method, the database occurrences and the fieldwork data, detailing the number of miners at each field site, were first separated based on the region to which they belong (table 7). It was necessary to perform the analysis at the regional scale, as the intensity and scale of mining activities varies from region to region. For example, over the past 3 years, mining activity has decreased in the Kindia-Télimélé region and has increased in

Table 6. Results for estimating the volume of gravel dug per person per day in Guinea.

[m², square meter; m, meter; m³, cubic meter]

Site	Mean area of pit (m²)	Mean over-burden thickness (m)	Mean gravel thickness (m)	Total thickness of pit (m)	Percentage of gravel in pit	Volume of pit (m³)	Days to extract	Number of miners per pit	Total volume of material dug per person per day (m³)	Total volume of gravel dug per person per day (m³)
Bouroumaya	25	6	0.5	6.5	0.08	162.5	20	5	1.63	0.13
Banyama	4	2	0.5	2.5	0.2	10	6	3	0.56	0.11
Gbomilo	20	3	0.4	3.4	0.12	68	12	4	1.42	0.17
Safoulé	16	1.5	0.75	2.25	0.33	36	20	3	0.60	0.20
Kenenday North	25	4	0.5	4.5	0.11	112.5	30	3	1.25	0.14
Kignéfou	16	4.5	0.5	5	0.10	80	24	3	1.11	0.11
Average										**0.14**

Table 7. Results of the Averaging Method calculations for estimating the number of miners per region in Guinea.

Region	Recorded database occurrences	Activity level (percent)	Estimated number of active sites	Mean number of miners per field site	Mean number of miners per region
Forécariah-Coyah	34	75	26	266	6,789
Kindia-Télimélé	22	30	7	18	119
Kissidougou-Macenta	268	60	161	220	35,376
Kérouané	235	60	141	336	47,336
Other	100	60	60	105	6,300
Total					**95,920**

the Forécariah-Coyah region. Meanwhile, mining activities have remained fairly stable in the Kissidougou-Macenta and Kérouané regions, as these are highly mineralized regions. It was necessary to ensure therefore that such differences are represented in the calculations. Next, the field sites within each region that contained the minimum and maximum number of miners were removed, as these values were considered to be outliers and are not accurate representations of the dataset. The minimum number of miners at a site is zero, if the site is inactive, or one, as is the case when a single person digs a pit to prospect for diamonds. Sites in the prospecting stage are not considered in this study to be active sites; they may become an active site if diamonds are discovered and mining expands. Therefore, such minimum values within each region were removed from the analysis. The largest sites within each region were also removed. For example, in the Kérouané region, the largest field site, had 5,500 miners—3,200 more miners than the second largest site in the region. This particular site was a former industrial mining site, operated by AREDOR, and is a well-known site with a long history. It

is not, however, an accurate reflection of activity within the region as a whole, as evidenced by the much lower number of miners at the remaining field sites in the region. Therefore, to arrive at an average number of miners per field site for each region, the minimum and maximum values were removed, and an average number of miners per field site was calculated for each region. Next, the number of active sites within each region was determined by multiplying the number of database occurrences within each region by a predetermined percentage, based on the intensity of mining activities in the region as a whole. This step was necessary because a large proportion of the database occurrences are historical records, indicative of previous mining activity which may have since ceased, or they simply mark isolated occurrences that never became active mining sites. In Forécariah-Coyah, a region in which activity has recently increased, it was estimated that 75 percent of the database occurrences were active mine sites, while in Kindia-Télimélé, a region which has experienced a recent decline in activity, it was estimated that 30 percent of the occurrences were active. In Kissidougou-Macenta and

Kérouané, both regions of relatively stable activity levels, it was estimated that 60 percent of the database occurrences represent active sites. To arrive at the final number of miners per region following this method, the number of active database occurrences was multiplied by the average number of miners per field site. The resulting values were then totaled to calculate an estimate of the total number of diamond miners in Guinea.

The second method, known as the Quartile Method, uses the same database occurrence and field site data, though the method arrives at the total number of miners through a slightly different approach (table 8). The first several steps are similar. The database occurrences and the field site data were separated by region, and the field sites containing the minimum and maximum values were removed for each region. The database occurrences were then multiplied by the same percentages used in the Averaging Method calculations in order to estimate the number of active mine sites. The remaining field sites were then divided equally into four quartile ranges, from which an average number of miners was calculated for each quartile.

Table 8. Results of the Quartile Method calculations for estimating the number of miners per region in Guinea.

Quartile range	Average number of miners	Number of active sites	Total
Forécariah-Coyah			
9–256.75	133	6	797
256.75–504.5	381	6	2,284
504.5–752.25	628	6	3,770
752.25–1,000	876	6	5,257
Total			**12,108**
Kindia-Télimélé			
9–14.25	12	2	23
14.25–19.5	17	2	34
19.5–24.75	22	2	44
24.75–30	27	2	55
Total			**156**
Kissidougou-Macenta			
30–172.5	101	40	4,050
172.5–315	244	40	9,750
315–457.5	386	40	15,450
457.5–600	529	40	21,150
Total			**50,400**
Kérouané			
80–210	145	35	5,075
210–340	275	35	9,625
340–470	405	35	14,175
470–600	535	35	18,725
Total			**47,600**
Total number of miners			**110,264**

The database occurrences were separated equally among the quartiles. Field data confirm that not all sites have the same number of workers. In order to account for variations in the scale of operations, an equal number of occurrences was assigned to each quartile range. The average number of miners per quartile was then multiplied by the number of database occurrences per quartile. These calculations were performed for each of the four regions. The total number of miners for each region was summed to arrive at a second estimate of the total number of diamond miners in Guinea.

GIS Modeling Results

Diamond Resource Potential Results

Field mapping and soil profile information were used to assist with modeling the diamond occurrences in the overall watersheds. The estimated grade, gravel thickness, and area calculations (table 6) were done for alluvial flats and terraces in the diamondiferous watersheds. The western zone diamond deposits in the Forécariah-Coyah and Kindia-Télimélé Prefectures fall within the Atlantic drainage basins as part of the Konkouré, Kolenté, Forécariah, and Malikouré Rivers. Diamond deposits in Kissidougou-Macenta and Kérouané occur in both the Niger basin as well as river basins draining into the Atlantic. Analyzing the diamond deposits based on their regions and drainage basins more accurately reflects the fluvial geomorphology parameters of each of the deposits (fig. 6).

The estimated grade and gravel thicknesses were developed from records within the diamond occurrence database and from fieldwork (table 9). The surface areas of the alluvial flats and terraces were calculated from the landform terrain modeling done in the GIS. An example of the results of the landform modeling of the Mandala River subbasin is shown in figure 32. The modeling results show the alluvial flat and terrace areas for all of the rivers and streams. However, to analyze the diamond resource potential, only the segments of rivers or streams containing a diamond occurrence were used in the estimation. The volume of diamondiferous gravel is calculated by multiplying the average gravel thickness by the surface area of the alluvial flats and terraces. Three-quarters of the volume of gravel is multiplied by the "basic" grade, while the remaining quarter of the volume is multiplied by the higher "concentration" grade. The above-mentioned variables were then entered into equation 1 to calculate the resource potential for each of the regional subbasins, the results of which were then totaled to develop an overall diamond resource potential.

The diamond resource potential of the Forécariah-Coyah region is estimated to be approximately 1.3 million carats. The Kindia-Télimélé bedrock fracture deposits are estimated to contain 1.4 million carats. In southeastern Guinea, the Kissidougou-Macenta and Kérouané regions are estimated to contain about 20 million carats each. Finally, the Other

Table 9. Results of the diamond resource potential evaluation for the five regions studied in Guinea.

[m², square meter; m, meter; m³, cubic meter; kt/m³, carats per cubic meter]

Region	Number of cells	Cell area (m²)	Total surface area (m²)	Gravel thickness (m)	Total volume of deposit (m³)	Volume of basic grade deposits (m³)	Volume of concentration grade deposits (m³)	Estimated basic grade (kt/m³)	Estimated concentration grade (kt/m³)	Basic grade reserves (kt)	Concentration grade reserves (kt)
Forécariah-Coyah											
Alluvial flat (AF)	3,214	8,100	26,033,400	0.58	15,099,372	11,324,529	3,774,843	0.05	0.25	566,226.45	943,710.75
Alluvial terrace (AT)	4,942	8,100	40,030,200	0.25	10,007,550	7,505,662.5	2,501,887.5	0.025	0.15	187,641.56	375,283.13
Kindia-Télimélé											
Bedrock fractures	1,566	8,100	12,684,600	1.27	16,109,442	12,082,081.5	4,027,360.5	0.05	0.35	604,104.08	1,409,576.18
Kissidougou-Macenta											
Alluvial flat (AF)	32,064	8,100	259,718,400	0.52	135,053,568	101,290,176	33,763,392	0.05	0.5	5,064,508.80	16,881,696.00
Alluvial terrace (AT)	27,139	8,100	219,825,900	0.25	54,956,475	41,217,356.25	13,739,118.75	0.025	0.25	1,030,433.91	3,434,779.69
Kérouané											
Alluvial flat (AF)	17,910	8,100	145,071,000	0.36	52,225,560	39,169,170	13,056,390	0.05	1	1,958,458.50	13,056,390.00
Alluvial terrace (AT)	26,155	8,100	211,855,500	0.25	52,963,875	39,722,906.25	13,240,968.75	0.025	0.5	993,072.66	6,620,484.38
Other											
Alluvial flat (AF)	9,086	8,100	73,596,600	0.5	36,798,300	27,598,725	9,199,575	0.05	0.35	1,379,936.25	3,219,851.25
Alluvial terrace (AT)	9,815	8,100	79,501,500	0.25	19,875,375	14,906,531.25	4,968,843.75	0.025	0.15	372,663.28	745,326.56
Subtotal										12,157,045.48	46,687,097.93
Total											58,844,143.41
Previously exploited diamonds											20,000,000
Final total											38,844,143.41

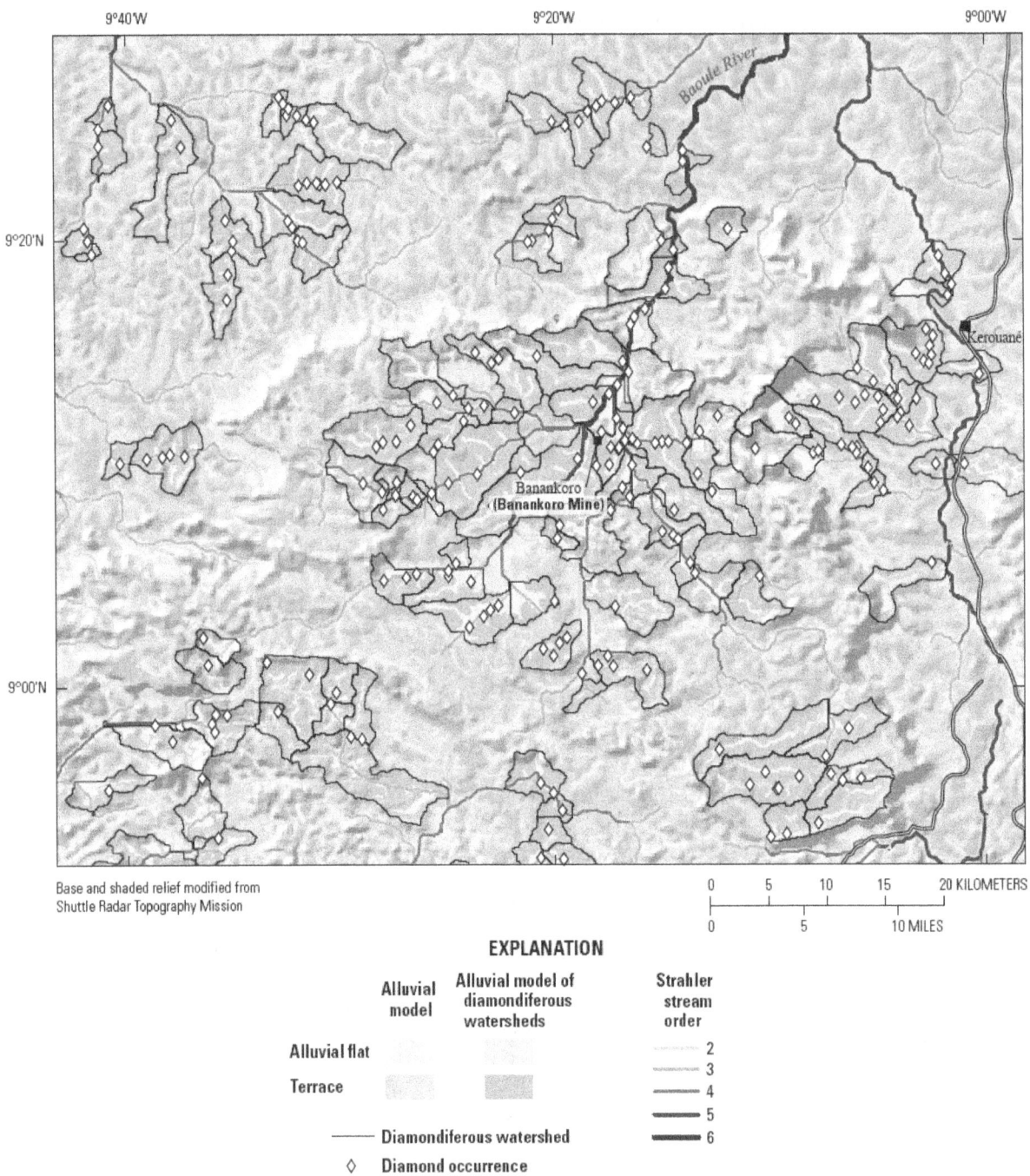

Figure 32. Results of the landform modeling of the Mandala River subbasin showing occurrence of alluvial flats and terraces mapped for the Baoulé River.

region is estimated to contain approximately 4 million carats. The total amount of estimated diamond reserves for all of the regions is approximately 60 million carats. However, the number of diamonds already exploited must be subtracted from this total to calculate the current remaining resource potential. Based on available data, the number of carats that have already been mined in Guinea is estimated to be between 15 and 20 million. It must be noted that currently there are two gaps in diamond-production statistics, the first gap extending from the years 1949 through 1953, and the second gap extending from the years 1962 through 1989. Once subtracted from the total estimated reserves, the revised total diamond resource is approximately 40 million carats. This estimate is 10–15 million carats greater than the previous estimate completed by Morgan and others (1992), in which 25–30 million carats were estimated to be remaining in Guinea. This previous estimate, however, was only of Guinea's southeastern diamond mining prefectures and did not take into account the country's western deposits. The estimate calculated in this study is based on all current mining activities taking place in the country and therefore is the most comprehensive estimate of Guinea's diamond resources that has been produced to date.

Diamond-Production Capacity Results

The diamond-production capacity was estimated using equation 2. The average volume of the gravel takes into account patchy low-grade deposits and the fact that many diggers mine zones barren of diamonds because they lack the equipment to sample the zones. The volume of diamond-iferous gravel mined per day was estimated to be 0.14 m³ for all regions, while the average grade of the deposits was esti-mated to vary regionally from 0.08 to 0.2 kt/m³. The number of days worked per year is estimated to be 216, which takes into

account the general absence of mining activities on Fridays and during the rainy season, which typically occurs from August through October. The total number of artisanal miners is estimated to be 103,000. Using the Averaging Method, it was calculated that there are approximately 95,000 artisanal diamond miners in Guinea, while using the Quartile Method it was calculated that there are approximately 110,000 artisanal diamond miners in Guinea. The results of the two methods were averaged, to arrive at an estimated 103,000 artisanal diamond miners. Finally, total industrial production was estimated to be 100,000 kt, based on available mining company production numbers. The total production capacity, calculated for the five regions, was estimated to be approximately 600,000 kt. Production capacity is lowest in Kindia-Télimélé (426 kt), followed by the "Other" region (16,670 kt), followed by Forécariah-Coyah (26,139 kt), Kissidougou-Macenta (145,545 kt), and Kérouané (417,655 kt) (table 10). This trend supports field observations, which suggest that mining activity has been declining substantially in Kindia-Télimélé and increasing in Forécariah-Coyah, with many miners moving from the former region to the latter. Kérouané has traditionally been Guinea's most intensively mined prefecture and is the location of large artisanal and industrial mining operations; therefore, it is natural that Kérouané has the highest production capacity of the five regions.

Multiple variables, the values of which were calculated from historical records and field data, were utilized to arrive at the production capacity estimates. It is possible that the values of one or more of the variables used in the equation may overestimate or underestimate the situation on the ground. However, due to the significant amount of data collected during the multiple fieldwork missions and the exhaustive background literature review, the total production capacity likely does not fall below or above 20 percent of the estimated 600,000 kt. A realistic range of production for Guinea there-fore is 480,000 to 720,000 kt.

Table 10. Results of the production capacity assessment per region.

[m³, cubic meter; kt/m³, carats per cubic meter; kt, carat]

Region	Volume of material worked per digger per day (Vm/d) (m³)	Average gravel grade (g) (kt/m³)	Total number of days a digger works per year (d)	Estimated number of diggers (A_i)	Industrial production (Ip) (kt)	Current production capacity (P_i) (kt)
Forécariah-Coyah	0.14	0.08	247	9,449	0	26,139
Kindia-Télimélé	0.14	0.13	177	137	0	426
Kissidougou-Macenta	0.14	0.12	202	42,888	0	145,545
Kérouané	0.14	0.20	239	47,468	100,000	417,655
Other	0.14	0.09	216	6,300	0	16,670
Total						**606,434**

Conclusion

The goal of this study was to estimate the alluvial diamond resource endowment and the current production capacity of the alluvial diamond mining sector of Guinea. A modified volume and grade methodology was used to estimate the remaining diamond reserves within five diamondiferous regions, while the diamond-production capacity of these regions was estimated by inputting the number of artisanal miners, the number of days artisans work per year, and the average grade of the deposits into a formulaic expression.

Guinea's resource potential was estimated to be 40 million carats, while the production capacity was estimated to lie within a range of 480,000 to 720,000 carats per year. While preliminary results have been produced based on the integration of historical documents, four fieldwork campaigns, and remote sensing and GIS analysis, data gaps remain. In particular, this study would benefit from the incorporation of additional grade estimates of diamond deposits throughout the main diamond mining regions. Such data would either confirm the estimated grades used in the assessment, thus validating the results, or could potentially require a new, updated calculation of production.

Over the past 6 years, production in Guinea has twice exceeded the estimated 480,000–720,000 carat production capacity and has twice fallen below this estimate. The unusually high production figures reported in 2007 and 2008 led to the passing of the Administrative Decision on Guinea and an assessment of the deposits to determine a realistic production range. According to the results of this assessment, the 1,018,723 carats and 3,098,490 carats produced in 2007 and 2008, respectively, fall well outside of the estimated range. The Guinean government has since worked with the KP to strengthen internal controls on their diamond exports, and production figures have been significantly reduced. In fact, in 2010 and 2011, production fell below the estimated range by approximately 100,000 carats. It is important to note that the estimate produced in this study is attainable only given the best possible conditions. However, many external factors can cause conditions to be less than ideal, including a lack of infrastructure in remote diamond mining areas and the mining of low-grade, non-economical deposits. Furthermore, artisanal miners are a transitory population, and the number of miners in the diamond sector may fluctuate significantly based on commodity prices. For example, the steady increase in the price of gold over the past decade has led to many miners leaving the diamond sector for the gold sector. Artisanal diamond mining is a highly dynamic sector, and it is therefore challenging to make firm production calculations. The results estimated in this study, however, were developed through an examination of all available information concerning Guinea's diamond deposits, and therefore is the best available guide to assessing Guinea's diamond deposits. While data gaps remain, this assessment provides a reliable baseline from which other studies may be conducted. The summary of the information here is also a step toward greater information transparency with respect to alluvial diamond resources.

The cooperation that took place between the USGS, Guinean government, and civil society organizations was a crucial component to the data collection process. The training in field methods and data collection, conducted by the USGS in 2011 for members of the MMG and civil society organizations, was the first step in this process. As a result of the training, members of the Guinean government and civil society were able to successfully implement fieldwork missions in two additional, heavily mined prefectures in southeastern Guinea, as well as several more active sites in western Guinea. The acquisition of data from these additional sites provided the most comprehensive overview of the current situation of mining activities in Guinea to date. However, artisanal diamond mining is a dynamic sector, with miners moving in and out of due to fluctuations in market prices and resource discovery and (or) depletion. Therefore, the continued monitoring and evaluation of the sector is required to address and account for the changes that will inevitably occur. The Guinean government now has the tools and experience to spearhead the monitoring process and ensure that their diamond exports meet the standards outline by the KP.

References Cited

Bardet, M.G., 1974, Géologie du diamants—Gisements de diamants d'Afrique: Orléans, Bureau de Recherches Géologiques et Minières Memoir, chap. 21.

Barthélémy, Francis, and others, 2006, Republic of the Congo, diamond potential, production capacity, and the Kimberley Process—Final report: Bureau de Recherches Géologiques et Minières, RC-54589-EN, 99 p.

Bering, D., Brinckmann, J., Camara, N'Doungou, Diawara, Mahmoud, Gast, Lothar, and Kieita, Sékou, 1998, Evaluation de l'Inventaire des Ressources Minérales de Guinée: Hannover, Bundesanstalt für Geowissenschaften und Rohstoffe.

Bermúdez-Lugo, Omayra, 2004, The mineral industry of Guinea: USGS Minerals Yearbook, 5 p.

Chirico, P.G., Barthélémy, Francis, and Koné, Fatiaga, 2010a, Alluvial diamond resource potential and production capacity assessment of Mali: U.S. Geological Survey Scientific Investigations Report 2010–5044, 23 p.

Chirico, P.G., Barthélémy, Francis, and Ngbokoto, F.A., 2010b, Alluvial diamond resource potential and production capacity assessment of the Central African Republic: U.S. Geological Survey Scientific Investigations Report 2010–5043, 22 p.

Chirico, P.G., Malpeli, K.C., Anum, Solomon, and Phillips, E.C., 2010c, Alluvial diamond resource potential and production capacity assessment of Ghana: U.S. Geological Survey Scientific Investigations Report 2010–5045, 25 p.

Ellis, Roger, 1987, Aredor makes the grade: Mining Magazine, v. 157, no. 3, p. 206–213.

Greenhalgh, P.A.L., 1985, West African diamonds, 1919–1983—An economic history: Manchester, Manchester University Press, 306 p.

Izon, David, 1994, The mineral industry of Guinea: USGS Minerals Yearbook, 4 p.

Janse, A.J.A, 1996, A history of diamond sources in Africa—Part II: Gems and gemology, v. 32, no. 1, 30 p.

Kimberley Process, 2004–2011, Kimberley Process Rough Diamond Statistics, available at *http://www.kimberleyprocess. com/documents/transparency_statistics_en.html*.

Kimberley Process Certification Scheme Secretariat (KPCS), 2009, Kimberley Process Plenary Session Communique : Kimberley Process Certification Scheme Secretariat, available at *http://www.kimberleyprocess.com/documents/ plenary_intersessional_meeting_en.html*.

Marshall, T.R., and Baxter-Brown, R., 1995, Basic principles of alluvial diamond exploration: Journal of Geochemical Exploration, v. 53, p. 277–292.

Ministère des Mines de l'Energie et de l'Hydraulique (MMEH), 2009, Arrete A/2009/4130 MMEH/SGG/Portant autorisation de l'exploitation artisanale et de la commer-cialisation du diamanat et autres gemmes dans certaines prefectures: Republic of Guinea.

Morgan, G.A., Izon, David, and Ousamane Sow, Nene, 1992, The mineral economy of Guinea: U.S. Bureau of Mines, 24 p.

Moyar, Antoine, and Buxtant, E., 1963, The diamond industry in 1960–1961: Antwerp, Vlaams Economisch Verbond, 112 p.

Property Rights and Artisanal Development Pilot Program (PRADD), 2008, Policy review—The artisanal diamond mining sector in the Republic of Guinea: U.S. Agency for International Development, 28 p.

Republic of Guinea, 1995, Mining code: Republic of Guinea, 53 p., available at *http://us-africa.tripod.com/ consulaatguinee/photos/Mining-Code.pdf*.

Rombouts, Luc, 1987a, Geology and evaluation of the Guinean diamond deposits: Annales de la Société géolo-gique de Belgique, v. 110, p. 241–259.

Rombouts, Luc, 1987b, Evaluation of low grade/high value diamond deposits: Mining Magazine, v. 157, no. 3, p. 217–220.

Rombouts, Luc, 1995, Sampling and statistical evaluation of diamond deposits: Journal of Geochemical Exploration, v. 53, p. 351–367.

Schlüter, Thomas, and Trauth, M.H., 2008, Geological atlas of Africa—With notes on stratigraphy, tectonics, economic geology, geohazards, geosites and geoscientific education of each country (2d ed.): Berlin, Springer, 307 p.

Sichel, H.S., 1973, Statistical valuation of diamondiferous deposits: Journal of the South African Institute of Mining and Metallurgy, p. 235–245.

Stellar Diamonds Limited, 2010, West African Diamonds PLC, Proposed acquisition, 211 p.

Strahler, A.N., 1964, Quantitative geomorphology of drainage basins and channel networks, *in* Chow, V.T., ed., Handbook of applied hydrology: McGraw-Hill, p. 4–40.

Sutherland, D.G., 1993, Drainage basin evolution in Southeast Guinea and the development of diamondiferous placer deposits: Economic Geology, v. 88, p. 44–54.

Sutherland, D.G., and Robinson, A.D., 1996, Characteristics of alluvial diamond deposits of the River Sarabaya, SE Guinea: Africa Geoscience Review, v. 3, no. 2, p. 317–329.

Swiecki, Rafal, 2008, Diamonds in Africa—Republic of Guinea: Alluvial Exploration and Mining Web site, 2 p., *http://www.minelinks.com/alluvial/diamondGeology60.html*.

Teeuw, R.M., Thomas, M.F., and Thorp, M.B., 1991, Alluvial mining: Institution of Mining and Metallurgy, p. 458–480.

USAID, 2012, Property Rights and Artisanal Diamond Development (PRADD) training to combat illicit diamond trafficking in West Africa: Data collection in Guinea: USAID, 21 p.

Walker, P.W.A., and Sobie, P.A., 2008, National instrument 43-101 technical report prepared on the diamond mining and exploration properties in Guinea, West Africa for Stellar Diamonds Limited, Panmore Gordon (UK) Limited and GMP Securities Europe LLP.

Wright, J.B., Hastings, D.A., Jones, W.B., and Williams, H.R., 1985, Geology and mineral resources of West Africa: London, George Allen and UnWin, 187 p.

www.ingramcontent.com/pod-product-compliance
Lightning Source LLC
Chambersburg PA
CBHW081618170526
45166CB00009B/3015